CRISPR

Ziheng Zhang · Ping Wang · Ji-Long Liu
Editors

CRISPR

A Machine-Generated Literature Overview

 Springer

Editors
Ziheng Zhang
School of Life Science and Technology
ShanghaiTech University
Shanghai, China

Ping Wang
University Library
ShanghaiTech University
Shanghai, China

Ji-Long Liu ⓘ
School of Life Science and Technology
ShanghaiTech University
Shanghai, China

ISBN 978-981-16-8503-3   ISBN 978-981-16-8504-0 (eBook)
https://doi.org/10.1007/978-981-16-8504-0

This Springer imprint is published by the registered company Springer Nature Singapore Pte Ltd.
The registered company address is: 152 Beach Road, #21-01/04 Gateway East, Singapore 189721,
Singapore

Preface

Clustered regularly interspaced short palindromic repeats (CRISPR) is a continuously evolving technology, with the first report of a specific sequence in *Escherichia coli* published in 1987. In 2007, CRISPR was considered a bacterial immune system used to resist virus invasion. It was applied to gene editing in 2012. Since then, research in this area has increased exponentially. From a practical viewpoint, the possibility of assimilating all the research relating to CRISPR is diminishing. Therefore, we directed artificial intelligence (AI) to write this book to provide an overview of this huge field of research. We are pleased to be able to collaborate with Springer Nature in producing this book entitled, "*CRISPR—A Machine-Generated Literature Overview*".

Some issues for consideration are whether the source of literature involves copyright issues, the volume of literature and the size of the final book. The book, therefore, is divided it into several chapters. We think it is an interesting attempt at a first edition. We hope this book is helpful to readers. This book is the first AI book in biology published by Springer Nature. If you find any mistakes in it, please forgive us.

CRISPR has a variety of applications. Initially, it was used as a means of gene editing; however, CRISPR has many possible applications in cell biology, developmental biology and so on. CRISPR is a technology that allows protein, RNA and DNA to be combined precisely. Then, various combinations can be derived, such as protein modification, RNA modification and DNA modification, from which many technologies can be derived. This is helpful for basic research and provides ideas for clinical, transformation, industry and synthesis. This was also our motivation for using CRISPR as the theme of this AI book.

The enormous task of collating research in this area has been discussed at length. Initially, the number of studies published by Springer Nature directly relating to CRISPR topics was about 5000, which made the task of covering every single paper in one book impossible. Therefore, it became more urgent to capture representative studies that reflected important and promising topics. Mapping knowledge networks remarkably accelerates the design of "CRISPR—A Machine-Generated Literature Overview" and also inspires the classification of four sub-domain book chapters (in

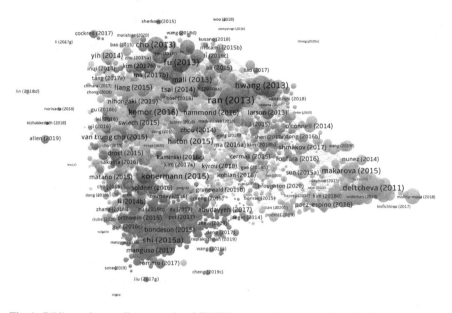

Fig. 1 Bibliography coupling networks of CRISPR-topic references published by Springer Nature

addition, Chaps. 1 and 6 are summaries of reviews and protocols). In this book, CRISPR-related studies were identified by bibliography coupling networks. Here, bibliographic coupling is an association between scientific publications when they have one or more references in common (Fig. 1).

First, all CRISPR-related scientific studies published by Springer Nature were obtained as original data for research network construction. Second, bibliography coupling networks and several clusters of subdomains illustrated by VOSviewer were generated to produce an overview of the research field. Subsequently, highly weighted nodes (papers) were collected. We then checked the validity of these papers and cluster results. Irrelevant nodes were eliminated and parameters, such as resolution and minimum size, of clusters were modified. Finally, after several iterations, the final targeted 114 studies were imported to the Dimensions AutoSummarization platform for book creation.

Many thanks to Dr. Mengchu Huang and Janina Krieger of Springer Nature, and the support of the platform and AI team. We hope readers can also give us valuable advice.

Shanghai, China   Ziheng Zhang
                                                                                 Ping Wang
                                                                                 Ji-long Liu

Contents

Chapter 1
Development and Vision of CRISPR-Based Technology

Ziheng Zhang, Ping Wang, and Ji-Long Liu

Machine generated keywords: crispr technology, researcher, epigenetic, chromatin, talen, zfns, desire, technology, field, review, alteration, future, biological, molecular, state.

CRISPR-Cas systems for editing, regulating and targeting genomes.

https://doi.org/10.1038/nbt.2842

Abstract-Summary

Genome editing mediated by these nucleases has been used to rapidly, easily and efficiently modify endogenous genes in a wide variety of biomedically important cell types and in organisms that have traditionally been challenging to manipulate genetically.

A modified version of the CRISPR-Cas9 system has been developed to recruit heterologous domains that can regulate endogenous gene expression or label specific genomic loci in living cells.

Although the genome-wide specificities of CRISPR-Cas9 systems remain to be fully defined, the power of these systems to perform targeted, highly efficient alterations of genome sequence and gene expression will undoubtedly transform biological research and spur the development of novel molecular therapeutics for human disease.

Z. Zhang
School of Life Science and Technology, ShanghaiTech University, Shanghai, China

P. Wang
University Library, ShanghaiTech University, Shanghai, China

J.-L. Liu (✉)
School of Life Science and Technology, ShanghaiTech University, Shanghai, China

1

Main

Strategies for efficiently inducing precise, targeted genome alterations were limited to certain organisms (e.g., homologous recombination in yeast or recombineering in mice) and often required drug-selectable markers or left behind 'scar' sequences associated with the modification method (e.g., residual loxP sites from Cre recombinase-mediated excision).

Targeted genome editing using customized nucleases provides a general method for inducing targeted deletions, insertions and precise sequence changes in a broad range of organisms and cell types.

Early methods for targeting DSB-inducing nucleases to specific genomic sites relied on protein-based systems with customizable DNA-binding specificities, such as meganucleases, zinc finger nucleases (ZFNs) and transcription activator–like effector nucleases (TALENs).

To ZFN and TALEN methods, which use protein-DNA interactions for targeting, RNA-guided nucleases (RGNs) use simple, base-pairing rules between an engineered RNA and the target DNA site.

ZFNs and TALENs are artificial fusion proteins composed of an engineered DNA binding domain fused to a nonspecific nuclease domain from the FokI restriction enzyme.

From a bacterial CRISPR immune system to engineered RGNs

The protospacer-encoded portion of the crRNA directs Cas9 to cleave complementary target-DNA sequences, if they are adjacent to short sequences known as protospacer adjacent motifs (PAMs).

The type II CRISPR system from S. pyogenes has been adapted for inducing sequence-specific DSBs and targeted genome editing [1].

With this system, Cas9 nuclease activity can be directed to any DNA sequence of the form N_{20}-NGG simply by altering the first 20 nt of the gRNA to correspond to the target DNA sequence.

Type II CRISPR systems from other species of bacteria that recognize alternative PAM sequences and that utilize different crRNA and tracrRNA sequences have also been used for targeted genome editing [2–4].

Cas9 variants that cut one strand rather than both strands of the target DNA site (known as 'nickases') have also been shown to be useful for genome editing.

Determining the specificities of RNA-guided Cas9 nucleases

To assess RGN specificity, several groups have created gRNA variants containing one to four nucleotide mismatches in the complementarity region and have then examined the abilities of these molecules to direct Cas9 nuclease activity in human cells at reporter gene (J.K.J. and colleagues [5]) or endogenous gene [6, 7] target sites.

A reciprocal, and perhaps more relevant, approach for studying specificity is to assess the activities of Cas9 at potential off-target genomic DNA target sites, (i.e., sites that have a few nucleotide differences compared to the intended target).

These results are similar to those of another study, which used in vitro selection for Cas9 nuclease cleavage activity to identify potential off-target sites from a partially degenerate library of target site variants.

It is worth noting that deep sequencing the genomes of individual cell clones is expected to be neither sensitive nor effective for defining the full genome-wide spectrum of Cas9 off-target sites because each clone would likely only carry mutations at a small proportion of, if any, possible off-target sites.

Methods for reducing off-target effects of Cas9 nucleases

To single Cas9 nickases (which can at some sites more favorably induce HDR events relative to NHEJ indels), paired Cas9 nickases targeted to sites on opposite DNA strands separated by 4 to 100 bp can efficiently introduce both indel mutations and HDR events with a single-stranded DNA oligonucleotide donor template in mammalian cells [8–10].

Paired nickases can reduce Cas9-induced off-target effects of gRNAs in human cells; the addition of a second gRNA and substitution of Cas9 nickase for Cas9 nuclease can lead to lower levels of unwanted mutations at previously known off-target sites of the original gRNA [8].

These truncated gRNAs (which we refer to as 'tru-gRNAs') have 17 or 18 nucleotides of complementarity; they generally function as efficiently as full-length gRNAs in directing on-target Cas9 activity but show decreased mutagenic effects at off-target sites and enhanced sensitivity to single or double mismatches at the gRNA:DNA interface [11].

Practical considerations for implementing CRISPR-Cas technology

The choice of promoter used to express gRNAs can limit the options for potential target DNA sites.

In cultured mammalian cells, researchers have used electroporation [12], nucleofection [5, 13] and Lipofectamine-mediated transfection [5, 6, 13] of nonreplicating plasmid DNA to transiently express Cas9 and gRNAs.

For most RGN applications, transient expression of gRNAs and Cas9 is typically sufficient to induce efficient genome editing.

Although constitutive expression of RGN components might potentially lead to higher on-target editing efficiencies, extended persistence of these components in the cell might also lead to increased frequencies of off-target mutations, a phenomenon that has been previously reported with ZFNs [14].

Presumably, each gRNA will be expected to have a different range of off-target effects and therefore if the same phenotype is observed with each of these different gRNAs it would seem unlikely that undesired mutations are the cause.

Applications of CRISPR-Cas beyond genome editing

Co-expression of the MS2 coat protein fusion with the hybrid gRNA and dCas9 has been used to recruit activation domains to a gene promoter in human cells [9].

Although the activation observed seems to be somewhat less robust than direct fusions to dCas9, this type of configuration might provide additional options and flexibility for recruitment of multiple effector domains to a promoter by, for example, using multiple gRNAs and MS2 coat protein binding sites on each gRNA to recruit many copies of different domains to the same promoter.

Evidence suggests that the effects of the small number of dCas9-activation or repression domain fusions tested to date can be highly specific in mammalian cells, as judged by RNA-seq or expression microarray experiments [15, 16]; however, this may be because not all binding events lead to changes in gene transcription.

Future directions

Methods for expanding the targeting range of RNA-guided Cas9 will be important for inducing precise HDR or NHEJ events as well as for implementing multiplex strategies, including paired nickases.

The field urgently needs to develop unbiased strategies to globally assess the off-target effects of Cas9 nucleases or paired nickases in any genome of interest.

Although tru-gRNAs and paired nickases can reduce off-target effects, it is likely that further improvements will be needed, especially for therapeutic applications.

The construction of inducible forms of Cas9 and/or gRNAs might provide a means to regulate the active concentration of these reagents in the cell and thereby improve the ratio of on- and off-target effects.

A related challenge will be to develop methods that enable expression of either the gRNAs or the Cas9 nuclease that is specific to a tissue, cell type or developmental stage.

Acknowledgements

A machine generated summary based on the work of Sander, Jeffry D; Joung, J Keith 2014 in Nature Biotechnology.

Applications of CRISPR technologies in research and beyond.

https://doi.org/10.1038/nbt.3659

Main

CRISPR-mediated genome editing could also expedite the development of large animal models of human diseases, including in primates, and thereby accelerate the identification of suitable therapies [17], although mosaicism issues have to be addressed to ensure the genotype of interest is consistently generated across tissues and cell types.

Before CRISPR-based gene therapies can be tested in human clinical trials, several practical issues and technical challenges need to be overcome, including the following: first, setting and reaching targets of accuracy and efficiency of both cleavage and repair at the cell population level; second, achieving efficient delivery to particular cell types, tissues or organs; third, understanding how to control various repair pathways; and fourth, predictably defining the mutational outcome of the DNA repair after DSB genesis.

Screen-based drug discovery approaches, together with the ability to use RNA-programmed genome editing technology to produce disease-recapitulating cell line models and animals, will continue to identify potential therapeutic targets.

Acknowledgements

A machine generated summary based on the work of Barrangou, Rodolphe; Doudna, Jennifer A 2016 in Nature Biotechnology.

The CRISPR tool kit for genome editing and beyond.

https://doi.org/10.1038/s41467-018-04252-2

Abstract-Summary

CRISPR is becoming an indispensable tool in biological research.

CRISPR-Cas9 is no longer just a gene-editing tool; the application areas of catalytically impaired inactive Cas9, including gene regulation, epigenetic editing, chromatin engineering, and imaging, now exceed the gene-editing functionality of WT Cas9.

We will present a brief history of gene-editing tools and describe the wide range of CRISPR-based genome-targeting tools.

Extended:

It will briefly discuss current and future impacts of these tools in science, medicine, and biotechnology.

Introduction

Even so-called serendipitous discoveries come when an inquisitive and open-minded researcher designs a series of careful experiments to follow an interesting observation.

During this process, researchers with creative minds and deep background knowledge can seize the opportunity to converge seemingly separate research fields and make a bigger scientific impact.

The genome-editing technologies and CRISPR tools have come to the current exciting stage through years of basic science research and progress from a large number of researchers.

This review will present the brief history and key developments in the field of genome editing and major genome-engineering tools.

Brief history of genome-editing efforts

Genomes of eukaryotic organisms are composed of billions of DNA bases.

Although such efforts drove a number of discoveries in molecular biology and genetics, the ability to precisely alter DNA in living eukaryotic cells came a few decades later.

Their studies demonstrated that mammalian cells can incorporate an exogenous copy of DNA into their own genome through a process called homologous recombination [18–20].

The rate of spontaneous integration of an exogenous DNA copy was extremely low (1 in 10^3–10^9 cells) [20].

And most critically, the approach could result in random integration of the exogenous copy into undesired genomic loci at a frequency similar to or higher than that of the target site [21].

Development of targeted nucleases for genome editing

To overcome such challenges, researchers started to re-engineer naturally existing meganucleases to alter their DNA-targeting specificities [22–24].

Unlike meganucleases, multiple zinc finger modules could be assembled into a larger complex to achieve higher DNA binding specificity.

The structure of zinc fingers was revealed, researchers started to create programmable nuclease proteins by fusing zinc finger proteins with the DNA cleavage domain of the Fok I endonuclease [25].

Knowing this, researchers removed the DNA sequence recognition domain of Fok I and fused only the DNA cleavage domain to zinc finger protein modules.

Designing two separate zinc finger modules that target two proximal sites next to each other allows Fok I to homodimerize and result in DNA strand breaks at the target sites.

Since each zinc finger recognized a 3-bp DNA code, combinatorial assembly of 6–7 zinc fingers out of the unique 64-finger pool (4^3 combinations) could uniquely target any 18–21 bp genomic sequence [26].

The rise of CRISPR as the genome-editing technology

The CRISPR gene-editing technology, as we know it today, is composed of an endonuclease protein whose DNA-targeting specificity and cutting activity can be programmed by a short guide RNA.

Unlike typical tandem repeats in the genome, the CRISPR repeat clusters were separated by non-repeating DNA sequences called spacers.

The computational analysis of these genomic sequences led researchers to notice key features of CRISPR repeat and spacer elements.

The spacer sequences of CRISPR dictate the targeting specificity of Cas enzymes, which provide defense against the phage [27].

Within a year after this key discovery, it was shown that the activity of Cas enzymes is guided by short CRISPR RNAs (crRNA) transcribed from the spacer sequences [28] and that it can block horizontal DNA transfer from bacterial plasmids [29].

The crucial work, which arguably marked the beginning of CRISPR as a biotechnology tool, has been the demonstration that Cas9 enzymes can be reprogrammed to target a desired DNA sequence in bacteria [1, 4].

Different CRISPR systems and their uses in genome editing

Class 2 contains type II, IV, V, and VI CRISPR systems [30].

Although researchers repurposed many different CRISPR/Cas systems for genome targeting, the most widely used one is the type II CRISPR-Cas9 system from Streptococcus pyogenes.

Because of the simple NGG PAM sequence requirements, S. pyogenes' Cas9 (spCas9) is used in many different applications.

Researchers are still actively exploring other CRISPR systems to identify Cas9-like effector proteins that may have differences in their sizes, PAM requirements, and substrate preferences.

Found Cas9 variants are large proteins, which adds particular limitation when it comes to their packaging and delivery into different cell types via Lenti or Adeno Associated viruses (AAV).

The tradeoff is that these smaller Cas9 proteins require more complex PAM sequences.

These smaller Cas9 proteins have relatively limited targeting scope and flexibility in genome targeting compared to SpCas9 despite the reduction in size.

Re-engineering CRISPR-Cas9 tools

To these studies that expand the targeting scope of CRISPR tools, researchers are actively developing novel ways to increase the targeting specificity of the CRISPR-Cas9 system.

To these initial studies, researchers utilized alternative genome-wide tools to understand CRISPR-Cas9 targeting specificity.

To these approaches to assess the off-target effects of the system, several forward steps have been taken to increase the targeting specificity of CRISPR-Cas9 systems by re-engineering the existing spCas9 variants.

To such re-engineering efforts on the Cas9 structure, researchers are utilizing alternative targeting approaches to substantially reduce the off-target binding and cleavage activity of Cas9.

One of the easiest ways to increase the targeting specificity is changing the delivery method of the Cas9-sgRNA complex.

To such engineering approaches at the Cas9 protein, efforts also focused on modifying the sgRNA scaffold to increase the targeting specificity.

Utilizing CRISPR-Cas9 beyond genome editing

The review has focused on the basic mechanism of CRISPR targeting and some of the recent approaches that have been utilized to monitor or improve the targeting specificity of CRISPR-Cas9.

CRISPR-Cas9 has two catalytic domains (HNH and RuvC) that act together to mediate DNA DSBs [31].

Evolution of second-generation CRISPR gene-editing tools

Unlike WT Cas9, which results in DSBs and random indels at the target sites, these so-called second-generation genome-editing tools are able to precisely convert a single base into another without causing DNA DSBs.

The nickase Cas9 is the foundational platform for the base editor tools that enables direct C to T or A to G conversion at the target site without DSBs [32–34].

Komor and others recently demonstrated that a fusion complex composed of nickase Cas9 fused to an APOBEC1 deaminase enzyme and Uracyl Glycosylase inhibitor (UGI) protein effectively converts Cytosine (C) into Thymine (T) at the target site without causing double strand DNA breaks [33].

A transfer RNA adenosine deaminase has also been evolved and fused to nickase Cas9 to develop another novel base editor that achieves direct A–G conversion at the target sites [32].

For further details about various applications of CRISPR base-editing tools, please refer to review articles that comprehensively cover these novel application areas.

CRISPR-mediated gene expression regulation

This has been exploited to develop the CRISPR interference (CRISPRi) approach in which dCas9 binding activity blocks the transcriptional process and thus knocks down (KD) gene expression [35].

Fusing a strong repressor complex such as Kruppel-associated Box (KRAB) to dCas9 results in a stronger and more specific gene repressor than dCas9 alone [16].

These transcriptional regulators further recruit additional co-repressor proteins such as KRAB-box-associated protein-1 (KAP-1) and epigenetic readers such as heterochromatin protein 1 (HP1) proteins to repress genes [36].

To dCas9-KRAB-mediated gene repression, using the dCas9-targeting platform to recruit strong transcriptional activators results in robust induction of gene expression.

In one such study, researchers used the engineered sgRNA-MS2 scaffold to recruit MCP-fused VP64 [37] or the P65-HSF1 transactivation complex (HSF1: heat shock transcription activator) [38] to activate expression from an endogenous locus.

Called a synergistic activation mediator (SAM) complex, in addition to dCas9-VP64 fusion complex, MCP-fused P65-HSF1 transactivation domains were recruited to the target site through the engineered sgRNA scaffold [38].

CRISPR-mediated epigenome editing

Although these epigenomic maps revealed unprecedented insight into cell-type-specific gene regulation and genome organization, the functional roles of various epigenomic features, such as histone modifications and DNA methylation, remain to be fully understood.

To deposit DNA methylation at a specifically targeted locus, researchers fused dCas9 to the catalytic domain of eukaryotic DNA methyl transferase (DNMT3A) [39–46] or prokaryotic DNA methyltransferase (MQ3) [43].

These early proof of principle studies have reported highly specific deposition of DNA methylation at the target loci and local effects on gene expression.

To targeted DNA methylation, active removal of local methylation marks from endogenous loci is another strategy to manipulate gene expression through DNA methylation.

Although robust locus-specific DNA demethylation and altered gene expression on target sites were reported, it remains to be seen whether dCas9-TET fusions may leave a global demethylation footprint akin to the methylation footprint of dCas9-fused methyltransferase.

CRISPR-mediated live cell chromatin imaging

Researchers used zinc fingers (ZNF) [47] and TALE proteins [48] for targeted recruitment of fluorescent proteins to repetitive genomic regions, such as centromeres and telomeres for live cell imaging.

Researchers used fluorescently labeled dCas9 to target repetitive regions of the genome to achieve the goal [49].

Transfection of as many as 26–36 unique sgRNAs is typically required to achieve live cell imaging of a non-repeat genomic region [49, 50].

To overcome this challenge, we recently utilized engineered sgRNA scaffolds which contains up to 16 MS2 binding modules to enable robust fluorescent signal amplification and allow imaging a repeat genomic region with as few as 4 sgRNAs [51].

CRISPR-mediated manipulation of chromatin topology

Targeted engineering of artificial chromatin loops between regulatory genomic regions provides a means to manipulate endogenous chromatin structures to understand their function and contribution to gene expression.

The demonstration that gene expression can be induced from a developmentally silent endogenous locus through forced chromatin looping was a significant step forward in demonstrating the potential for this system [52].

Researchers are now using dCas9-based platforms to achieve targeted and robust manipulation of chromatin structure and DNA loop formation.

Tethering these protein-dimerization systems, to two separate dCas9 orthologous, enabled forced chromatin loop formation between distal enhancer and promoter regions.

These proof of principle studies demonstrate the power of CRISPR as a targeted chromatin structure-rewiring tool.

Large-scale genetic and epigenetic CRISPR screenings

In such applications, instead of using a single sgRNA, WT Cas9 or dCas9-effector fusion proteins are guided with hundreds or thousands of individual sgRNAs in a population of cells.

In its basic form, a large pool of Cas9/sgRNAs are typically delivered to a population of cells via a low multiplicity of viral infection (MOI = 0.3 to 0.4).

The basic logic behind the CRISPR KO screenings is that if a gene is essential for a given phenotype, such as cell proliferation, then the cells infected with the sgRNAs targeting that gene will be relatively depleted from the population over time.

The relative abundance of each sgRNA in a given population of cells can be quantified by targeted sequencing.

Future directions

Surely, CRISPR-based technologies have empowered researchers with an unprecedented toolbox.

The history of molecular biology will place CRISPR-Cas9 among the major tools that enabled breakthrough discoveries and methodological advancements in science.

Several multiple recent review articles have comprehensively overviewed the specific applications of CRISPR tools [53–60].

The therapeutic applications of the CRISPR technologies are particularly exciting [61].

The rapid development of CRISPR-based tools also brings forth a number of technical challenges along with social and ethical concerns.

As CRISPR technologies grow in scope and power, social and ethical concerns over their use are also rising, and applications of these powerful tools deserve greater considerations [62].

The CRISPR-based technologies will undoubtedly continue to transform basic as well as clinical and biotechnological research.

One such obstacle is the potential immunogenicity to CRISPR-Cas9 proteins.

Acknowledgements

A machine generated summary based on the work of Adli, Mazhar 2018 in Nature Communications.

Genome editing with CRISPR–Cas nucleases, base editors, transposases and prime editors.

https://doi.org/10.1038/s41587-020-0561-9

Abstract-Summary

The development of new CRISPR–Cas genome editing tools continues to drive major advances in the life sciences.

Each tool comes with its own capabilities and limitations, and major efforts have broadened their editing capabilities, expanded their targeting scope and improved editing specificity.

We analyze key considerations when choosing genome editing agents and identify opportunities for future improvements and applications in basic research and therapeutics.

Main

The first demonstrations of programmable DNA cleavage by Cas9 nuclease [1, 4] and subsequent early demonstrations of its ability to carry out targeted genome modification in living eukaryotic cells [6, 13, 63–65] initiated an explosive growth in the discovery, engineering and application of CRISPR–Cas genome editing tools.

We restrict our discussion to the targeted alteration of genomic DNA sequence using CRISPR-based tools and refer readers to excellent reviews of topics related to other CRISPR applications such as transcriptional regulation [66–70], epigenetic modifications [67–70], RNA editing [69, 70] and nucleic acid detection [70].

The goal of a genome editing experiment is to convert a targeted DNA sequence into a new, desired DNA sequence (or sequences) in the native context of a cell's genome.

We discuss the development and application of base editors, genome-editing agents that precisely install point mutations without requiring double-stranded DNA breaks (DSBs) or donor DNA templates.

We summarize emerging CRISPR–Cas genome editing tools, including Cas transposons and recombinases, which mediate rearrangements of large segments of DNA, and prime editors, which directly copy edited sequences into target DNA sites in a manner that replaces the original DNA sequence.

Genome editing with CRISPR–Cas nucleases

We overview naturally occurring Cas effectors, the ways in which Cas nucleases have been used for genome editing, and the development of engineered Cas nuclease variants that enable broader targeting scope and higher DNA cleavage specificity.

SpCas9, currently the most widely used CRISPR–Cas nuclease, contains 1368 amino acids, recognizes a relatively common NGG PAM, can be used with either an sgRNA or crRNA/tracrRNA pair, functions optimally with 20-nt spacers, has robust DNA targeting and cleavage activity, and supports relatively high levels of off-target editing [69, 71, 72].

Nureki and co-workers [73] used structure-guided rational design to develop SpCas9-NG, a Cas9 variant that can target all NG PAM sequences with varying activities, in many cases with higher efficiency than xCas9-3.7.

An extensive comparison of SpCas9, xCas9-3.7 and SpCas9-NG activity in human cells [74] revealed that nucleotides outside of the traditional trinucleotide PAM sequence, extending to up to 5 nucleotides adjacent to the protospacer, can influence targeting activity by each of these Cas9 variants.

Genome editing with base editors

The rapid increase in the number of known CBE- and ABE-compatible Cas domains with distinct PAM preferences, truncation or extension of the sgRNA to alter editing outcomes [75, 76], as well as the increase in deaminase domains with different target sequence compatibilities and editing window positions, greatly increases the probability that a base editor exists that can convert a given target nucleotide of interest.

Base editors can also induce Cas-independent DNA off-target editing, which occurs from the long-term expression of deaminases that can randomly deaminate transiently accessible nucleotides at a low level across the genome.

Highly sequence-specific or narrow-window deaminases can be combined with the ever-growing set of Cas domains to create editors with targeting breadth, efficiency and DNA specificity properties that are well-suited for most base editing applications [77, 78].

The deaminase domains of CBEs and ABEs, which are derived from enzymes that natively deaminate RNA, can also induce Cas-independent off-target RNA base editing when overexpressed at high levels in mammalian cells [79–82].

Transposases and recombinases

Described natural CRISPR-associated transposases and engineered Cas-domain-fused transposase systems can integrate genomic cargos in vitro and into bacterial genomes [83–85].

Engineered Cas-fused recombinases, which can in theory insert, delete, invert or replace target DNA, have also been reported to modify plasmid substrates in mammalian cells and to delete targeted genomic DNA in human cells, albeit so far with low efficiency and substantial target sequence restrictions [86].

Although the sequence constraints of known recombinase domains, including Ginβ, limit their use, the ability of recombinases to perform a wide diversity of genome-modifying activities including insertions, deletions, replacements and inversions make them exciting targets for continued development.

Identifying, engineering and further developing CRISPR-targeted transposases and recombinases represent exciting opportunities in genome editing that may enable precise rearrangements of large DNA sequences of interest.

Prime editing

Prime editors are able to install point mutations at distances far (>30 bp) from the site of Cas9 nicking, which offers greater targeting flexibility than nuclease-mediated HDR with ssDNA donor templates, which typically are unable to introduce edits efficiently more than ~ 10 bp from the cut site [87, 88].

Prime editors can mediate all types of local edits, however, including those inaccessible to base editors; may be more easily targeted than other precision CRISPR editing methods because of greater flexibility in the distance between the PAM and the edit; and also appear to be free of bystander editing since the sequence of the RT template determines the sequence of the edited DNA.

These issues include illuminating the cell-state or cell-type determinants of prime editing efficiency, understanding the DNA repair mechanisms that result in productive or unproductive prime editing, and developing delivery strategies for in vivo applications that require delivery of the prime editor protein and pegRNA.

Conclusions

The rapid evolution of genome-editing technologies has led us into a new era—an era in which we can now edit our own genomes, as well as the genomes of many other organisms that affect our communities.

Continued efforts to improve editing capabilities, to understand all the consequences of editing our genomes, to innovate new ways to deliver editing agents into cells, and to fully engage scientists, doctors, ethicists, governments and other stakeholders will be crucial to guide our next steps and to ensure that these scientific advances can realize their full potential to benefit society.

Acknowledgements

A machine generated summary based on the work of Anzalone, Andrew V.; Koblan, Luke W.; Liu, David R. 2020 in Nature Biotechnology.

CRISPR technologies for precise epigenome editing.

https://doi.org/10.1038/s41556-020-00620-7

Abstract-Summary

The epigenome involves a complex set of cellular processes governing genomic activity.

Dissecting this complexity necessitates the development of tools capable of specifically manipulating these processes.

With future optimization, CRISPR-based epigenomic editing stands as a set of powerful tools for understanding and controlling biological function.

Introduction

Efforts to understand the source of this diversity has given rise to the field of epigenetics [89], which encompasses a wide variety of factors beyond the DNA sequence that regulate genomic function.

These factors in combination form an 'epigenetic state' of the cell.

While techniques such as biochemical in vitro reconstitution of chromatin [90–92], disruption of epigenetic effectors [93–95] or manipulation of DNA regulatory elements and other genomic features [93, 96] have revealed the contributions of epigenetic states, obtaining a comprehensive, detailed picture requires technologies that can induce precise, defined epigenetic changes at specific loci in living cells.

Targeted epigenetic changes can be achieved via tethering epigenetic effectors to sequence-specific DNA-binding domains (DBDs).

This localizes the epigenetic alteration near the DBD-bound site, thereby establishing causal links between effector presence, induced epigenetic changes and gene regulation.

The use of DBDs has formed a critical foundation for targeted epigenetic engineering [40, 93, 94, 97–103].

Chromatin editing meets the CRISPR revolution

dCas recruitment of epigenetic effectors has been reported to lead to the following locus-specific editing of chromatin marks: methylation at H3K4 (H3K4me; for example, by PRDM9, LSD1, SMYD3 and BAF) [104–107], H3K9me (for example, by heterochromatin protein 1 (HP1), G9A and KRAB) [40, 105, 107–110], H3K27me (for example, by EZH2 and FOG1) [40, 105, 107, 110] and H3K79me (by DOT1L) [104]; acetylation at H3K27 (H3K27ac; for example, by p300/CBP and HDAC3) [110–121] and other histone residues [114]; and DNA methylation (for example, by DNA methyltransferase 3A (DNMT3A), M.SssI and TET1) [39–42, 44–46, 102, 107, 119, 122–139].

Combining these modulators with the wide-ranging deployability of dCas was used to create high-density maps of how alterations in chromatin marks within a specific DNA region affected gene expression [39, 96, 105, 108, 112, 113, 115, 116, 128, 137–142].

One important facet of the epigenetics field is manipulating gene expression through chromatin edits to rationally engineer cellular behaviour.

An inability to achieve large changes in gene expression can reflect both experimental limitations in editing an epigenetic state and inherent biological effects of chromatin editing.

Questions such as whether a particular mark has fundamental limits on its ability to affect expression, whether a mark can affect expression of any gene or is limited to a subset of contexts and whether mark modifications are effective in the presence of endogenous epigenetic programmes remain unanswered, which all point to the continuing necessity for deeper exploration via epigenetic engineering.

CRISPR-mediated engineering of the diverse epigenome

dCas-based approaches provide complementary information, for instance, by perturbing long ncRNA expression via CRISPRi/a [143–147] or by targeting epigenetic edits (that is, histone methylation [105, 108, 112, 137–142, 148], histone acetylation [112, 113, 115, 116], DNA methylation [39, 128, 148] or direct CRISPRa [149]) at various points along DNA-regulatory elements.

dCas-based chromatin editing sometimes demonstrated lower changes at the dCas-binding site [43, 45, 136], which suggests that dCas may also modulate chromatin editing by protecting its binding site from epigenetic modifications.

dCas can be used to study the causal effect of genome organization by directly mobilizing the bound DNA element.

Using dCas to simultaneously recruit an exogenously provided enhancer element increased CRISPRa activity [150], which again highlights the potential of combinatorial epigenetic engineering.

Other studies used dCas to directly alter CTCF binding at a specific locus by direct occlusion [148] or DNA methylation editing [39, 43, 124, 148], thereby demonstrating the interlinks between chromatin state, CTCF-driven topology changes and gene expression.

The recruitment of TF domains involved in phase separation via dCas enabled examination of the biophysics of condensates and their interaction with chromatin within cells in a locus-specific manner [151].

The frontier of CRISPR epigenome editing

The rapidly growing toolset of inducible controllers of CRISPR technologies, such as chemical [113, 114, 117, 152–154], light [152, 155, 156] and protein-based controllers [128, 154, 156], will facilitate investigations into the dependence of epigenetic editing and maintenance on the timescale of induction.

A chemically inducible CRISPR system produced either temporary or stable epigenetic changes depending on the duration of induced perturbation [153], which highlights the importance of this avenue of study.

CRISPR systems may facilitate the development of epigenetic measurement techniques.

CRISPR technologies will therefore expand the toolkit to not only perturb but also measure epigenetic states within the cell.

CRISPR-based epigenetic engineering could be applied to treat additional diseases with epigenetic mechanisms [157].

Counteracting aberrant epigenetic changes has been proven as a therapeutically useful strategy in cancer [158], and it is possible that CRISPR epigenetic approaches may target a wider range of epigenetic diseases.

Further investigations will further open up the potential of CRISPR epigenome-editing technologies to treat and prevent diseases.

Outlook

Many fundamental questions regarding how epigenetics causally alter cellular function across the diverse range of (epi-)genomic contexts may finally be answered.

This advancement in knowledge will provide novel cellular engineering modalities, allowing for the co-option of existing epigenetic pathways in the cell to unlock applications involving stable, wide-ranging, inheritable and reversible changes in gene expression.

Acknowledgements

A machine generated summary based on the work of Nakamura, Muneaki; Gao, Yuchen; Dominguez, Antonia A.; Qi, Lei S 2021 in Nature Cell Biology.

References

1. Jinek, M., et al. 2012. A programmable dual-RNA-guided DNA endonuclease in adaptive bacterial immunity. *Science* 337: 816–821.
2. Esvelt, K.M., et al. 2013. Orthogonal Cas9 proteins for RNA-guided gene regulation and editing. *Nature Methods* 10: 1116–1121.
3. Hou, Z., et al. 2013. Efficient genome engineering in human pluripotent stem cells using Cas9 from Neisseria meningitidis. *Proceedings of the National academy of Sciences of the United States of America* 110: 15644–15649.
4. Gasiunas, G., R. Barrangou, P. Horvath, and V. Siksnys. 2012. Cas9-crRNA ribonucleoprotein complex mediates specific DNA cleavage for adaptive immunity in bacteria. *Proceedings of the National academy of Sciences of the United States of America* 109: E2579–E2586.
5. Fu, Y., et al. 2013. High-frequency off-target mutagenesis induced by CRISPR-Cas nucleases in human cells. *Nature Biotechnology* 31: 822–826.
6. Cong, L., et al. 2013. Multiplex genome engineering using CRISPR/Cas systems. *Science* 339: 819–823.
7. Hsu, P.D., et al. 2013. DNA targeting specificity of RNA-guided Cas9 nucleases. *Nature Biotechnology* 31: 827–832.

8. Ran, F.A., et al. 2013. Double nicking by RNA-guided CRISPR Cas9 for enhanced genome editing specificity. *Cell* 154: 1380–1389.
9. Mali, P., et al. 2013. CAS9 transcriptional activators for target specificity screening and paired nickases for cooperative genome engineering. *Nature Biotechnology* 31: 833–838.
10. Cho, S.W., et al. 2014. Analysis of off-target effects of CRISPR/Cas-derived RNA-guided endonucleases and nickases. *Genome Research* 24: 132–141.
11. Fu, Y., J.D. Sander, D. Reyon, V.M. Cascio, and J.K. Joung. 2014. Improving CRISPR-Cas nuclease specificity using truncated guide RNAs. *Nature Biotechnology*. https://doi.org/10.1038/nbt.2808(26January.
12. Ding, Q., et al. 2013. Enhanced efficiency of human pluripotent stem cell genome editing through replacing TALENs with CRISPRs. *Cell Stem Cell* 12: 393–394.
13. Mali, P., et al. 2013. RNA-guided human genome engineering via Cas9. *Science* 339: 823–826.
14. Gaj, T., J. Guo, Y. Kato, S.J. Sirk, and C.F. Barbas III. 2012. Targeted gene knockout by direct delivery of zinc-finger nuclease proteins. *Nature Methods* 9: 805–807.
15. Cheng, A.W., et al. 2013. Multiplexed activation of endogenous genes by CRISPR-on, an RNA-guided transcriptional activator system. *Cell Research* 23: 1163–1171.
16. Gilbert, L.A., et al. 2013. CRISPR-mediated modular RNA-guided regulation of transcription in eukaryotes. *Cell* 154: 442–451.
17. Niu, Y., et al. 2014. Generation of gene-modified cynomolgus monkey via Cas9/RNA-mediated gene targeting in one-cell embryos. *Cell* 156: 836–843.
18. Smithies, O., R.G. Gregg, S.S. Boggs, M.A. Koralewski, and R.S. Kucherlapati. 1985. Insertion of DNA sequences into the human chromosomal beta-globin locus by homologous recombination. *Nature* 317: 230–234.
19. Thomas, K.R., K.R. Folger, and M.R. Capecchi. 1986. High frequency targeting of genes to specific sites in the mammalian genome. *Cell* 44: 419–428.
20. Capecchi, M.R. 1989. Altering the genome by homologous recombination. *Science* 244: 1288–1292.
21. Lin, F.L., K. Sperle, and N. Sternberg. 1985. Recombination in mouse L cells between DNA introduced into cells and homologous chromosomal sequences. *Proceedings of the National academy of Sciences of the United States of America* 82: 1391–1395.
22. Sussman, D., et al. 2004. Isolation and characterization of new homing endonuclease specificities at individual target site positions. *Journal of Molecular Biology* 342: 31–41.
23. Seligman, L.M., et al. 2002. Mutations altering the cleavage specificity of a homing endonuclease. *Nucleic Acids Research* 30: 3870–3879.
24. Rosen, L.E., et al. 2006. Homing endonuclease I-CreI derivatives with novel DNA target specificities. *Nucleic Acids Research* 34: 4791–4800.
25. Kim, Y.G., J. Cha, and S. Chandrasegaran. 1996. Hybrid restriction enzymes: Zinc finger fusions to Fok I cleavage domain. *Proceedings of the National academy of Sciences of the United States of America* 93: 1156–1160.
26. Urnov, F.D., E.J. Rebar, M.C. Holmes, H.S. Zhang, and P.D. Gregory. 2010. Genome editing with engineered zinc finger nucleases. *Nature Reviews Genetics* 11: 636–646.
27. Barrangou, R., et al. 2007. CRISPR provides acquired resistance against viruses in prokaryotes. *Science* 315: 1709–1712.
28. Brouns, S.J., et al. 2008. Small CRISPR RNAs guide antiviral defense in prokaryotes. *Science* 321: 960–964.
29. Marraffini, L.A., and E.J. Sontheimer. 2008. CRISPR interference limits horizontal gene transfer in staphylococci by targeting DNA. *Science* 322: 1843–1845.
30. Koonin, E.V., K.S. Makarova, and F. Zhang. 2017. Diversity, classification and evolution of CRISPR-Cas systems. *Current Opinion in Microbiology* 37: 67–78.
31. Nishimasu, H., et al. 2014. Crystal structure of Cas9 in complex with guide RNA and target DNA. *Cell* 156: 935–949.
32. Gaudelli, N.M., et al. 2017. Programmable base editing of A*T to G*C in genomic DNA without DNA cleavage. *Nature* 551: 464–471.

33. Komor, A.C., Y.B. Kim, M.S. Packer, J.A. Zuris, and D.R. Liu. 2016. Programmable editing of a target base in genomic DNA without double-stranded DNA cleavage. *Nature* 533: 420–424.
34. Nishida, K. et al. 2016. Targeted nucleotide editing using hybrid prokaryotic and vertebrate adaptive immune systems. *Science* 353. https://doi.org/10.1126/science.aaf8729.
35. Qi, L.S., et al. 2013. Repurposing CRISPR as an RNA-guided platform for sequence-specific control of gene expression. *Cell* 152: 1173–1183.
36. Friedman, J.R., et al. 1996. KAP-1, a novel corepressor for the highly conserved KRAB repression domain. *Genes & Development* 10: 2067–2078.
37. Zalatan, J.G., et al. 2015. Engineering complex synthetic transcriptional programs with CRISPR RNA scaffolds. *Cell* 160: 339–350.
38. Konermann, S., et al. 2015. Genome-scale transcriptional activation by an engineered CRISPR-Cas9 complex. *Nature* 517: 583–588.
39. Liu, X.S., et al. 2016. Editing DNA methylation in the mammalian genome. *Cell* 167: 233–247.e17.
40. Amabile, A., et al. 2016. Inheritable silencing of endogenous genes by hit-and-run targeted epigenetic editing. *Cell* 167: 219-232.e214.
41. Vojta, A., et al. 2016. Repurposing the CRISPR-Cas9 system for targeted DNA methylation. *Nucleic Acids Research* 44: 5615–5628.
42. McDonald, J.I., et al. 2016. Reprogrammable CRISPR/Cas9-based system for inducing site-specific DNA methylation. *Biology Open* 5: 866–874.
43. Lei, Y., et al. 2017. Targeted DNA methylation in vivo using an engineered dCas9-MQ1 fusion protein. *Nature Communications* 8: 16026.
44. Xiong, T., et al. 2017. Targeted DNA methylation in human cells using engineered dCas9-methyltransferases. *Science and Reports* 7: 6732.
45. Morita, S., et al. 2016. Targeted DNA demethylation in vivo using dCas9-peptide repeat and scFv-TET1 catalytic domain fusions. *Nature Biotechnology* 34: 1060–1065.
46. Xu, X., et al. 2016. A CRISPR-based approach for targeted DNA demethylation. *Cell Discovery* 2: 16009.
47. Lindhout, B.I. et al. 2007. Live cell imaging of repetitive DNA sequences via GFP-tagged polydactyl zinc finger proteins. *Nucleic Acids Research* 35: e107.
48. Miyanari, Y., C. Ziegler-Birling, and M.E. Torres-Padilla. 2013. Live visualization of chromatin dynamics with fluorescent TALEs. *Nature Structural & Molecular Biology* 20: 1321–1324.
49. Chen, B., et al. 2013. Dynamic imaging of genomic loci in living human cells by an optimized CRISPR/Cas system. *Cell* 155: 1479–1491.
50. Anton, T., S. Bultmann, H. Leonhardt, and Y. Markaki. 2014. Visualization of specific DNA sequences in living mouse embryonic stem cells with a programmable fluorescent CRISPR/Cas system. *Nucleus* 5: 163–172.
51. Qin, P., et al. 2017. Live cell imaging of low- and non-repetitive chromosome loci using CRISPR-Cas9. *Nature Communications* 8: 14725.
52. Deng, W., et al. 2014. Reactivation of developmentally silenced globin genes by forced chromatin looping. *Cell* 158: 849–860.
53. Doench, J.G. 2018. Am I ready for CRISPR? A user's guide to genetic screens. *Nature Reviews Genetics* 19: 67–80.
54. Hsu, P.D., E.S. Lander, and F. Zhang. 2014. Development and applications of CRISPR–Cas9 for genome engineering. *Cell* 157: 1262–1278.
55. Pulecio, J., N. Verma, E. Mejia-Ramirez, D. Huangfu, and A. Raya. 2017. CRISPR/Cas9-based engineering of the epigenome. *Cell Stem Cell* 21: 431–447.
56. Barrangou, R., and P. Horvath. 2017. A decade of discovery: CRISPR functions and applications. *Nature Microbiology* 2: 17092.
57. Sander, J.D., and J.K. Joung. 2014. CRISPR-Cas systems for editing, regulating and targeting genomes. *Nature Biotechnology* 32: 347–355.
58. Doudna, J.A., and E. Charpentier. 2014. Genome editing. The new frontier of genome engineering with CRISPR-Cas9. *Science* 346: 1258096.

59. Fellmann, C., B.G. Gowen, P.C. Lin, J.A. Doudna, and J.E. Corn. 2017. Cornerstones of CRISPR-Cas in drug discovery and therapy. *Nature Reviews. Drug Discovery* 16: 89–100.

60. Wang, H., M. La Russa, and L.S. Qi. 2016. CRISPR/Cas9 in genome editing and beyond. *Annual Review of Biochemistry* 85: 227–264.

61. Dunbar, C.E. et al. 2018. Gene therapy comes of age. Science 359. https://doi.org/10.1126/science.aan4672.

62. Baltimore, D. et al. 2015. Biotechnology. A prudent path forward for genomic engineering and germline gene modification. *Science* 348: 36–38.

63. Hwang, W.Y., et al. 2013. Efficient genome editing in zebrafish using a CRISPR-Cas system. *Nature Biotechnology* 31: 227–229.

64. Cho, S.W., S. Kim, J.M. Kim, and J.-S. Kim. 2013. Targeted genome engineering in human cells with the Cas9 RNA-guided endonuclease. *Nature Biotechnology* 31: 230–232.

65. Jinek, M. et al. 2013. RNA-programmed genome editing in human cells. *Elife* 2: e00471.

66. Shalem, O., N.E. Sanjana, and F. Zhang. 2015. High-throughput functional genomics using CRISPR-Cas9. *Nature Reviews Genetics* 16: 299–311.

67. Dominguez, A.A., W.A. Lim, and L.S. Qi. 2016. Beyond editing: Repurposing CRISPR-Cas9 for precision genome regulation and interrogation. *Nature Reviews Molecular Cell Biology* 17: 5–15.

68. Thakore, P.I., J.B. Black, I.B. Hilton, and C.A. Gersbach. 2016. Editing the epigenome: Technologies for programmable transcription and epigenetic modulation. *Nature Methods* 13: 127–137.

69. Adli, M. 2018. The CRISPR tool kit for genome editing and beyond. *Nature Communications* 9: 1911.

70. Pickar-Oliver, A., and C.A. Gersbach. 2019. The next generation of CRISPR-Cas technologies and applications. *Nature Reviews Molecular Cell Biology* 20: 490–507.

71. Komor, A.C., A.H. Badran, and D.R. Liu. 2017. CRISPR-based technologies for the manipulation of eukaryotic genomes. *Cell* 168: 20–36.

72. Kim, D., K. Luk, S.A. Wolfe, and J.-S. Kim. 2019. Evaluating and enhancing target specificity of gene-editing nucleases and deaminases. *Annual Review of Biochemistry* 88: 191–220.

73. Nishimasu, H., et al. 2018. Engineered CRISPR-Cas9 nuclease with expanded targeting space. *Science* 361: 1259–1262.

74. Kim, H.K., et al. 2020. High-throughput analysis of the activities of xCas9, SpCas9-NG and SpCas9 at matched and mismatched target sequences in human cells. *Nature Biomed Engineering* 4: 111–124.

75. Lee, J.K., et al. 2018. Directed evolution of CRISPR-Cas9 to increase its specificity. *Nature Communications* 9: 3048.

76. Ryu, S.-M., et al. 2018. Adenine base editing in mouse embryos and an adult mouse model of Duchenne muscular dystrophy. *Nature Biotechnology* 36: 536–539.

77. Yu, Y., et al. 2020. Cytosine base editors with minimized unguided DNA and RNA off-target events and high on-target activity. *Nature Communications* 11: 2052.

78. Doman, J.L., A. Raguram, G.A. Newby, and D.R. Liu. 2020. Evaluation and minimization of Cas9-independent off-target DNA editing by cytosine base editors. *Nature Biotechnology* 38: 620–628.

79. Grünewald, J., et al. 2019. CRISPR DNA base editors with reduced RNA off-target and self-editing activities. *Nature Biotechnology* 37: 1041–1048.

80. Zhou, C., et al. 2019. Off-target RNA mutation induced by DNA base editing and its elimination by mutagenesis. *Nature* 571: 275–278.

81. Grünewald, J., et al. 2019. Transcriptome-wide off-target RNA editing induced by CRISPR-guided DNA base editors. *Nature* 569: 433–437.

82. Rees, H.A., C. Wilson, J.L. Doman, and D.R. Liu. 2019. Analysis and minimization of cellular RNA editing by DNA adenine base editors. Science Advances 5: eaax5717.

83. Klompe, S.E., P.L.H. Vo, T.S. Halpin-Healy, and S.H. Sternberg. 2019. Transposon-encoded CRISPR-Cas systems direct RNA-guided DNA integration. *Nature* 571: 219–225.

84. Strecker, J., et al. 2019. RNA-guided DNA insertion with CRISPR-associated transposases. *Science* 365: 48–53.
85. Chen, S.P., and H.H. Wang. 2019. An engineered Cas-transposon system for programmable and site-directed DNA transpositions. *CRISPR Journal* 2: 376–394.
86. Chaikind, B., J.L. Bessen, D.B. Thompson, J.H. Hu, and D.R. Liu. 2016. A programmable Cas9-serine recombinase fusion protein that operates on DNA sequences in mammalian cells. *Nucleic Acids Research* 44: 9758–9770.
87. Paquet, D., et al. 2016. Efficient introduction of specific homozygous and heterozygous mutations using CRISPR/Cas9. *Nature* 533: 125–129.
88. Liang, X., J. Potter, S. Kumar, N. Ravinder, and J.D. Chesnut. 2017. Enhanced CRISPR/Cas9-mediated precise genome editing by improved design and delivery of gRNA, Cas9 nuclease, and donor DNA. *Journal of Biotechnology* 241: 136–146.
89. Cavalli, G., and E. Heard. 2019. Advances in epigenetics link genetics to the environment and disease. *Nature* 571: 489–499.
90. Keung, A.J., J.K. Joung, A.S. Khalil, and J.J. Collins. 2015. Chromatin regulation at the frontier of synthetic biology. *Nature Reviews Genetics* 16: 159–171.
91. Sanulli, S., et al. 2019. HP1 reshapes nucleosome core to promote phase separation of heterochromatin. *Nature* 575: 390–394.
92. Wang, L., et al. 2019. Histone modifications regulate chromatin compartmentalization by contributing to a phase separation mechanism. *Molecular Cell* 76: 646-659.e6.
93. Stricker, S.H., A. Köferle, and S. Beck. 2017. From profiles to function in epigenomics. *Nature Reviews Genetics* 18: 51–66.
94. Holtzman, L., and C.A. Gersbach. 2018. Editing the epigenome: Reshaping the genomic landscape. *Annual Review of Genomics and Human Genetics* 19: 43–71.
95. Weinberg, D.N., et al. 2019. The histone mark H3K36me2 recruits DNMT3A and shapes the intergenic DNA methylation landscape. *Nature* 573: 281–286.
96. Gasperini, M., J.M. Tome, and J. Shendure. 2020. Towards a comprehensive catalogue of validated and target-linked human enhancers. *Nature Reviews Genetics* 21: 292–310.
97. Verschure, P.J., et al. 2005. In vivo HP1 targeting causes large-scale chromatin condensation and enhanced histone lysine methylation. *Molecular and Cellular Biology* 25: 4552–4564.
98. Hathaway, N.A., et al. 2012. Dynamics and memory of heterochromatin in living cells. *Cell* 149: 1447–1460.
99. Keung, A.J., C.J. Bashor, S. Kiriakov, J.J. Collins, and A.S. Khalil. 2014. Using targeted chromatin regulators to engineer combinatorial and spatial transcriptional regulation. *Cell* 158: 110–120.
100. Bintu, L., et al. 2016. Dynamics of epigenetic regulation at the single-cell level. *Science* 351: 720–724.
101. Gao, X. et al. 2014. Comparison of TALE designer transcription factors and the CRISPR/dCas9 in regulation of gene expression by targeting enhancers. *Nucleic Acids Research* 42: e155.
102. Yamazaki, T. et al. 2017. Targeted DNA methylation in pericentromeres with genome editing-based artificial DNA methyltransferase. *PLoS One* 12: e0177764.
103. Mlambo, T., et al. 2018. Designer epigenome modifiers enable robust and sustained gene silencing in clinically relevant human cells. *Nucleic Acids Research* 46: 4456–4468.
104. Cano-Rodriguez, D., et al. 2016. Writing of H3K4Me3 overcomes epigenetic silencing in a sustained but context-dependent manner. *Nature Communications* 7: 12284.
105. Kearns, N.A., et al. 2015. Functional annotation of native enhancers with a Cas9-histone demethylase fusion. *Nature Methods* 12: 401–403.
106. Kim, J.-M., et al. 2015. Cooperation between SMYD3 and PC4 drives a distinct transcriptional program in cancer cells. *Nucleic Acids Research* 43: 8868–8883.
107. Wang, H., et al. 2018. Epigenetic targeting of Granulin in hepatoma cells by synthetic CRISPR dCas9 Epi-suppressors. *Molecular Therapy Nucleic Acids* 11: 23–33.
108. Thakore, P.I., et al. 2015. Highly specific epigenome editing by CRISPR-Cas9 repressors for silencing of distal regulatory elements. *Nature Methods* 12: 1143–1149.

109. Braun, S.M.G., et al. 2017. Rapid and reversible epigenome editing by endogenous chromatin regulators. *Nature Communications* 8: 560.
110. Li, K., et al. 2020. Interrogation of enhancer function by enhancer-targeting CRISPR epigenetic editing. *Nature Communications* 11: 485.
111. Liu, P., M. Chen, Y. Liu, L.S. Qi, and S. Ding. 2018. CRISPR-based chromatin remodeling of the endogenous Oct4 or Sox2 locus enables reprogramming to pluripotency. *Cell Stem Cell* 22: 252-261.e4.
112. Klann, T.S., et al. 2017. CRISPR-Cas9 epigenome editing enables high-throughput screening for functional regulatory elements in the human genome. *Nature Biotechnology* 35: 561–568.
113. Kuscu, C., et al. 2019. Temporal and spatial epigenome editing allows precise gene regulation in mammalian cells. *Journal of Molecular Biology* 431: 111–121.
114. Chiarella, A.M., et al. 2020. Dose-dependent activation of gene expression is achieved using CRISPR and small molecules that recruit endogenous chromatin machinery. *Nature Biotechnology* 38: 50–55.
115. Hilton, I.B., et al. 2015. Epigenome editing by a CRISPR-Cas9-based acetyltransferase activates genes from promoters and enhancers. *Nature Biotechnology* 33: 510–517.
116. Cheng, A.W., et al. 2016. Casilio: A versatile CRISPR–Cas9-Pumilio hybrid for gene regulation and genomic labeling. *Cell Research* 26: 254–257.
117. Chen, T., et al. 2017. Chemically controlled epigenome editing through an inducible dCas9 system. *Journal of the American Chemical Society* 139: 11337–11340.
118. Kwon, D.Y., Y.T. Zhao, J.M. Lamonica, and Z. Zhou. 2017. Locus-specific histone deacetylation using a synthetic CRISPR-Cas9-based HDAC. *Nature Communications* 8: 15315.
119. Okada, M., M. Kanamori, K. Someya, H. Nakatsukasa, and A. Yoshimura. 2017. Stabilization of Foxp3 expression by CRISPR–dCas9-based epigenome editing in mouse primary T cells. *Epigenetics & Chromatin* 10: 24.
120. Shrimp, J.H., et al. 2018. Chemical control of a CRISPR–Cas9 acetyltransferase. *ACS Chemical Biology* 13: 455–460.
121. Zhang, X., et al. 2018. Gene activation in human cells using CRISPR/Cpf1-p300 and CRISPR/Cpf1-SunTag systems. *Protein & Cell* 9: 380–383.
122. Saunderson, E.A., et al. 2017. Hit-and-run epigenetic editing prevents senescence entry in primary breast cells from healthy donors. *Nature Communications* 8: 1450.
123. Galonska, C., et al. 2018. Genome-wide tracking of dCas9-methyltransferase footprints. *Nature Communications* 9: 597.
124. Pflueger, C., et al. 2018. A modular dCas9-SunTag DNMT3A epigenome editing system overcomes pervasive off-target activity of direct fusion dCas9–DNMT3A constructs. *Genome Research* 28: 1193–1206.
125. Lin, L., et al. 2018. Genome-wide determination of on-target and off-target characteristics for RNA-guided DNA methylation by dCas9 methyltransferases. *Gigascience* 7: 1–19.
126. Hofacker, D., et al. 2020. Engineering of effector domains for targeted DNA methylation with reduced off-target effects. *International Journal of Molecular Sciences* 21: 502.
127. Choudhury, S.R., Y. Cui, K. Lubecka, B. Stefanska, and J. Irudayaraj. 2016. CRISPR–dCas9 mediated TET1 targeting for selective DNA demethylation at BRCA1 promoter. *Oncotarget* 7: 46545–46556.
128. Liu, X.S., et al. 2018. Rescue of fragile X syndrome neurons by DNA methylation editing of the FMR1 gene. *Cell* 172: 979-992.e6.
129. Kantor, B., et al. 2018. Downregulation of SNCA expression by targeted editing of DNA methylation: A potential strategy for precision therapy in PD. *Molecular Therapy* 26: 2638–2649.
130. Marx, N., et al. 2018. CRISPR-based targeted epigenetic editing enables gene expression modulation of the silenced beta-galactoside alpha-2,6-sialyltransferase 1 in CHO cells. *Biotechnology Journal* 13: 1700217.

131. Mkannez, G., et al. 2018. DNA methylation of a PLPP3 MIR transposon-based enhancer promotes an osteogenic programme in calcific aortic valve disease. *Cardiovascular Research* 114: 1525–1535.

132. Xu, X., et al. 2018. High-fidelity CRISPR/Cas9-based gene-specific hydroxymethylation rescues gene expression and attenuates renal fibrosis. *Nature Communications* 9: 3509.

133. Ziller, M.J., et al. 2018. Dissecting the functional consequences of de novo DNA methylation dynamics in human motor neuron differentiation and physiology. *Cell Stem Cell* 22: 559-574.e9.

134. Baumann, V., et al. 2019. Targeted removal of epigenetic barriers during transcriptional reprogramming. *Nature Communications* 10: 2119.

135. Huang, Y.-H., et al. 2017. DNA epigenome editing using CRISPR–Cas SunTag-directed DNMT3A. *Genome Biology* 18: 176.

136. Stepper, P., et al. 2017. Efficient targeted DNA methylation with chimeric dCas9-Dnmt3a-Dnmt3L methyltransferase. *Nucleic Acids Research* 45: 1703–1713.

137. Fulco, C.P., et al. 2016. Systematic mapping of functional enhancer–promoter connections with CRISPR interference. *Science* 354: 769–773.

138. Xie, S., J. Duan, B. Li, P. Zhou, and G.C. Hon. 2017. Multiplexed engineering and analysis of combinatorial enhancer activity in single cells. *Molecular Cell* 66: 285-299.e5.

139. Morton, A.R., et al. 2019. Functional enhancers shape extrachromosomal oncogene amplifications. *Cell* 179: 1330-1341.e13.

140. Fulco, C.P., et al. 2019. Activity-by-contact model of enhancer–promoter regulation from thousands of CRISPR perturbations. *Nature Genetics* 51: 1664–1669.

141. Gasperini, M., et al. 2019. A genome-wide framework for mapping gene regulation via cellular genetic screens. *Cell* 176: 377-390.e19.

142. D'Ippolito, A.M., et al. 2018. Pre-established chromatin interactions mediate the genomic response to glucocorticoids. *Cell Systems* 7: 146-160.e7.

143. Raffeiner, P., et al. 2020. An MXD1-derived repressor peptide identifies noncoding mediators of MYC-driven cell proliferation. *Proceedings of the National academy of Sciences of the United States of America* 117: 6571–6579.

144. Zhu, S., et al. 2016. Genome-scale deletion screening of human long non-coding RNAs using a paired-guide RNA CRISPR–Cas9 library. *Nature Biotechnology* 34: 1279–1286.

145. Joung, J., et al. 2017. Genome-scale activation screen identifies a lncRNA locus regulating a gene neighbourhood. *Nature* 548: 343–346.

146. Liu, S.J. et al. 2017. CRISPRi-based genome-scale identification of functional long noncoding RNA loci in human cells. *Science* 355: eaah7111.

147. Bester, A.C., et al. 2018. An integrated genome-wide CRISPRa approach to functionalize lncRNAs in drug resistance. *Cell* 173: 649-664.e20.

148. Tarjan, D.R., W.A. Flavahan, and B.E. Bernstein. 2019. Epigenome editing strategies for the functional annotation of CTCF insulators. *Nature Communications* 10: 4258.

149. Simeonov, D.R., et al. 2017. Discovery of stimulation-responsive immune enhancers with CRISPR activation. *Nature* 549: 111–115.

150. Xu, X. et al. 2019. Gene activation by a CRISPR-assisted trans enhancer. *eLife* 8: e45973.

151. Shin, Y., et al. 2018. Liquid nuclear condensates mechanically sense and restructure the genome. *Cell* 175: 1481-1491.e13.

152. Gao, Y., et al. 2016. Complex transcriptional modulation with orthogonal and inducible dCas9 regulators. *Nature Methods* 13: 1043–1049.

153. Morgan, S.L., et al. 2017. Manipulation of nuclear architecture through CRISPR-mediated chromosomal looping. *Nature Communications* 8: 15993.

154. Nakamura, M., et al. 2019. Anti-CRISPR-mediated control of gene editing and synthetic circuits in eukaryotic cells. *Nature Communications* 10: 194.

155. Kim, J.H., et al. 2019. LADL: Light-activated dynamic looping for endogenous gene expression control. *Nature Methods* 16: 633–639.

156. Bubeck, F., et al. 2018. Engineered anti-CRISPR proteins for optogenetic control of CRISPR–Cas9. *Nature Methods* 15: 924–927.

157. Mirabella, A.C., B.M. Foster, and T. Bartke. 2016. Chromatin deregulation in disease. *Chromosoma* 125: 75–93.
158. Jones, P.A., J.-P.J. Issa, and S. Baylin. 2016. Targeting the cancer epigenome for therapy. *Nature Reviews Genetics* 17: 630–641.

Chapter 2
Gene Editing Through CRISPR-Based Technology

Ziheng Zhang, Ping Wang, and Ji-Long Liu

Introduction

Realizing arbitrary modification of DNA sequence is a major goal of genetic engineering and an important way for human beings to transform life. Initially, people used homologous recombination that relied on low frequency in nature to edit specific genes. This strategy was successfully applied to some lower organisms, such as yeast, and is still used today. For most organisms, however, the homologous recombination is low frequency in the absence of DNA double-strand breaks, especially for higher organisms, such as mammals. For this reason, a series of gene editing methods based on nucleases have been developed, i.e. simple and efficient gene editing tools, such as zinc finger nucleases (ZFN) and transcription activator-like effector nucleases (TALEN). The emergence of CRISPR generated a storm of gene editing, providing people with a simple and efficient approach.

CRISPR is derived from the immune system of bacteria and is used to prevent the invasion of bacteriophages. This feature of CRISPR is used to engineer a powerful gene editing system and for the application of gene editing of various species. At present, its skills have been evident in many fields, including gene function research, production of genetically modified animals, crop improvement and disease detection. It is easy to implement, and its high efficiency has greatly promoted the development of biology and medicine. At the same time, some risks have been reported, including its potential off-target and the ethical and moral aspects involved in using it for gene editing of human embryos. While enjoying the benefits of new technologies, attention should be paid to their risks and management. In this chapter, we introduce

Z. Zhang
School of Life Science and Technology, ShanghaiTech University, Shanghai, China

P. Wang
University Library, ShanghaiTech University, Shanghai, China

J.-L. Liu (✉)
School of Life Science and Technology, ShanghaiTech University, Shanghai, China

© The Author(s), under exclusive license to Springer Nature Singapore Pte Ltd. 2022
Z. Zhang et al. (eds.), *CRISPR*,
https://doi.org/10.1007/978-981-16-8504-0_2

some typical studies that used classic CRISPR research for gene editing, as well as some research discussing the potential risks of CRISPR.

Machine generated keywords: sgrna, type, immunity, zygote, phage, transcript, iii, gene expression, mouse, lineage, crrna, mrna, element, singlecell, infect.

2.1   CRISPR and The Bacterial Immunity

Machine generated keywords: phage, immunity, iii, infection, infect, type, host, bacteriophage, compartment, spacer, subtype, crrnas, strain, plasmid, virus.

The CRISPR/Cas bacterial immune system cleaves bacteriophage and plasmid DNA.

https://doi.org/10.1038/nature09523

Abstract-Summary

Clustered regularly interspaced short palindromic repeats (CRISPR) loci together with cas (CRISPR-associated) genes form the CRISPR/Cas immune system, which involves partially palindromic repeats separated by short stretches of DNA called spacers, acquired from extrachromosomal elements.

We show that the Streptococcus thermophilus CRISPR1/Cas system can also naturally acquire spacers from a self-replicating plasmid containing an antibiotic-resistance gene, leading to plasmid loss.

Acquired spacers that match antibiotic-resistance genes provide a novel means to naturally select bacteria that cannot uptake and disseminate such genes.

We also provide in vivo evidence that the CRISPR1/Cas system specifically cleaves plasmid and bacteriophage double-stranded DNA within the proto-spacer, at specific sites.

Main

The CRISPR/Cas immune systems act in at least two general steps: (1) the adaptation stage, where new spacers derived from foreign DNA (proto-spacers) are generally acquired at the leader end of the CRISPR locus [1, 2]; and (2) the interference stage, where the CRISPR/Cas system targets either invading DNA [3] or RNA [4].

The mechanistic details of spacer acquisition are still unknown, but a clearer picture is emerging for the interference stage, which starts with the transcription of the CRISPR locus from a promoter located within the leader sequence [5, 6].

We previously showed that when bacteriophage-sensitive Streptococcus thermophilus cells are infected by virulent bacteriophages, a subset of cells (frequency of $< 10^{-6}$) naturally diversify into bacteriophage-insensitive mutants through the acquisition of novel spacers derived from the invading bacteriophage genome into CRISPR1 and/or CRISPR3 [1, 7, 8].

We investigate the in vivo activity of the CRISPR/Cas system in S. thermophilus against both bacteriophage and plasmid DNA.

CRISPR/Cas affects plasmid stability

Plasmid stability assays were also performed using two isogenic DGCC7710 strains in which cas5 (csn1-like) or cas7 genes associated to CRISPR1 were inactivated before the introduction of pNT1.

Plasmid pNT1 was highly stable in the DGCC7710::pcas5⁻ mutant as no chloramphenicol-sensitive colonies could be isolated, out of 1,800 screened.

Of 170 randomly selected chloramphenicol-resistant colonies, none had acquired a new spacer in CRISPR1.

Chloramphenicol-sensitive colonies were readily obtained with the strain DGCC7710::pcas7⁻, but none of the 200 colonies tested had acquired a new spacer in CRISPR1, indicating that plasmid loss was probably the result of other mechanisms responsible for plasmid instability [9, 10].

These observations indicate that the CRISPR/Cas system causes plasmid loss in S. thermophilus.

CRISPR/Cas targets antibiotic-resistance genes

This tolerance for proto-spacer adjacent motif degeneracy could be due to the lower selective pressure for plasmids as compared to bacteriophages.

Some plasmid-interfering mutants targeting proto-spacers associated with non-consensus motifs (NNNAGAAG, NNATAAA, NNGGAAT or NNAGAAG) were also refractory to pNT1 reintroduction.

It is also worth mentioning that one plasmid-interfering mutant contained a spacer (S47) that matched the last 29 nucleotides (out of 30) of the corresponding proto-spacer in pNT1, indicating that sequence identity at the 5′ end of the spacer might be less important than in the middle or at the 3′ end.

Motif degeneracy influences plasmid interference

Although the NNAGTAG motif was initially permissive for pNT1, the CRISPR/Cas machinery still eliminated the circular and linear plasmid forms within a few generations.

To assess whether the observed plasmid linearization was the result of CRISPR/Cas activity, the cas5 and cas7 genes of PIM S46 were also inactivated and the isogenic strains transformed with pNT1.

We next investigated the terminal ends of the pNT1 DNA molecules by directly sequencing the linear plasmid extracted from the PIM S46 strain.

CRISPR/Cas system cuts viral DNA in the proto-spacers

We also investigated the fate of bacteriophage DNA during the infection of S. thermophilus BIM S4/S32, which contains two new spacers that target bacteriophage 2972.

Contrary to the method used for the determination of the cleavage site of plasmid pNT1, the low amount of cleaved bacteriophage DNA in the infected bacteriophage-insensitive mutants rendered direct sequencing of the extremities impossible.

The two cleavage sites within the S40 proto-spacer (a natural 5′-end truncated version of the S4 proto-spacer with 29 nucleotides instead of 30) were at the same position, indicating that the CRISPR1/Cas system in S. thermophilus probably acts in a 3′-end ruler-anchored manner.

Ligation and amplification of each cleaved fragment could be obtained and no nucleotide was missing in the S7, S41, S42 and S46 proto-spacer sequences, confirming the blunt-end cleavage activity of S. thermophilus CRISPR1/Cas system.

We have established that the S. thermophilus CRISPR1/Cas system cleaves both bacteriophage and plasmid DNA in vivo.

Methods Summary

S. thermophilus strains were grown at 37 °C or 42 °C in LM17 medium (ref 8).

S. thermophilus DGCC7710 (or its cas5⁻ and cas7⁻ derivatives) transformed with pNT1 (ref 11) was used to inoculate 10 ml of LM17.

One-hundred microlitres of the previous culture was inoculated into 10 ml of fresh LM17 medium every morning (42 °C) and night (37 °C) for 5 days, for a total of 9 inoculations.

PIM S55 and PIM S56 were also transformed with pLS1 (ref 12).

Plasmid content of re-transformed PIM S45 and PIM S46 was verified by extraction (Qiagen) and extremities of the linearized plasmid from PIM S46 were sequenced.

Online Methods

Streptococcus thermophilus strain DGCC7710 (ref 7) and bacteriophage-insensitive mutants (BIMs) or plasmid-interfering mutants (PIMs) were grown in M17 broth supplemented with 0.5% lactose (LM17) at 37 °C or 42 °C.

After removing a 1-ml uninfected sample, each bacterial culture was infected with purified bacteriophage 2972 at a multiplicity of infection (MOI) of 5, and incubated at 42 °C.

S. thermophilus DGCC7710 (or the cas5⁻ and cas7⁻ isogenic derivatives) transformed with pNT1 plasmid [11] (GenBank accession number HQ010044) was picked from a LM17 plate containing 5 μg ml⁻¹ of chloramphenicol.

The CRISPR1 of chloramphenicol-sensitive clones was verified: the 5′ end of the CRISPR1 of the plasmid-interfering mutants or bacteriophage-insensitive mutants was amplified by PCR with the primers yc70 (5′-TGCTGAGACAACCTAGTCTCTC-3′) [13] and RDS7revBamHI (5′-GGATCCGGATCCGTTGAGGCCTTGTTC-3′), and sequenced using the same primers.

All plasmid-interfering mutants (except for PIM S55 and PIM S56) could be transformed with the control vector pTRK687, which is carrying a chloramphenicol-resistance gene as selection marker.

Acknowledgements

Machine generated summary based on the work of Garneau, Josiane E.; Dupuis, Marie-Ève; Villion, Manuela; Romero, Dennis A.; Barrangou, Rodolphe; Boyaval, Patrick; Fremaux, Christophe; Horvath, Philippe; Magadán, Alfonso H.; Moineau, Sylvain, 2010 in Nature.

CRISPR RNA maturation by trans-encoded small RNA and host factor RNase III.

https://doi.org/10.1038/nature09886

Abstract-Summary

CRISPR/Cas systems constitute a widespread class of immunity systems that protect bacteria and archaea against phages and plasmids, and commonly use repeat/spacer-derived short crRNAs to silence foreign nucleic acids in a sequence-specific manner.

Although the maturation of crRNAs represents a key event in CRISPR activation, the responsible endoribonucleases (CasE, Cas6, Csy4) are missing in many CRISPR/Cas subtypes.

We show that tracrRNA directs the maturation of crRNAs by the activities of the widely conserved endogenous RNase III and the CRISPR-associated Csn1 protein; all these components are essential to protect S. pyogenes against prophage-derived DNA.

Main

In bacteria and archaea, CRISPR/Cas (clustered, regularly interspaced short palindromic repeats/CRISPR-associated proteins) constitutes an adaptive RNA-mediated defence system which targets invading phages or plasmids in three steps: (1) adaptation via integration of viral or plasmid DNA-derived spacers into the CRISPR locus, (2) expression of short guide CRISPR RNAs (crRNAs) consisting of unique single repeat-spacer units and (3) interference with the invading cognate foreign genomes by mechanisms that are yet to be fully understood [2–4, 7, 13–29].

Their homologues are missing in many CRISPR/Cas subtypes, suggesting the existence of alternate crRNA maturation pathways involving other Cas proteins and/or fundamentally different RNA processing events.

[Section 2]

We detected six crRNAs from CRISPR01 which were 39 to 42 nucleotides in length and probably were processed species, as judged by their depletion in the dRNA-seq library for primary transcripts.

tracrRNA directs pre-crRNA processing

The tracrRNA and pre-crRNA processing sites detected by dRNA-seq fell in the putative RNA duplex region, indicative of co-processing of the two RNAs upon pairing.

crRNA maturation requires RNase III and Csn1

According to our dRNA-seq data, the co-processed tracrRNA and pre-crRNA carry short 3' overhangs reminiscent of cleavage by the endoribonuclease RNase III [26, 30–34] or the related eukaryotic Dicer and Drosha enzymes [35–39].

Because none of the Cas proteins of CRISPR01 contains an RNase III-like motif, we hypothesized that the endogenous RNase III—a general RNA processing factor [30, 32, 40] encoded by the conserved rnc gene of the host—was recruited to cleave tracrRNA and pre-crRNA upon base pairing.

RNase III serves as a host factor in tracrRNA-mediated crRNA maturation, and constitutes the first example of a non-Cas protein that is recruited to CRISPR activity.

Our results show that in the absence of Cse3 (CasE), Cas6 or Csy4 proteins, CRISPR01 crRNA maturation is achieved by the concerted action of three novel factors, a trans-encoded small RNA, a host-encoded RNase and a Cas protein previously not implicated in pre-crRNA cleavage.

CRISPR immunity against prophage sequences

To investigate further the role of tracrRNA in CRISPR01-mediated immunity, we developed a plasmid-based read-out system that mimics infection with protospacer-containing lysogenic phages (a protospacer is a DNA target sequence that matches a CRISPR spacer).

tracrRNA homologues in CRISPR/Cas systems

RNA base-pairing might generally determine crRNA maturation in type II CRISPR/Cas systems, and based on RNA probing results, these systems seem to be constitutively activated to target and affect the maintenance of invader genomes.

The requirement of a trans-encoded small RNA for pre-crRNA processing into active crRNAs is a general RNA maturation mechanism shared by the type II (Nmeni /CASS4) CRISPR/Cas systems that lack the cse3 (casE), cas6 or csy4 gene but possess csn1.

Trans RNA-mediated activation of crRNA maturation to confer sequence-specific immunity against parasite genomes represents a novel RNA maturation pathway, and highlights the remarkable diversity and complexity of molecular mechanisms of CRISPR/Cas systems [2, 4, 15–19, 41, 42].

More studies are needed to determine whether an RNase III-mediated activation of a small effector RNA by co-processing with a trans-acting non-coding RNA is also used in other biological systems.

Methods Summary

Half of DNase I-treated SF370 total RNA was enriched for primary transcripts by treatment with Terminator 5'-phosphate-dependent exonuclease (TEX) (Epicentre), which degrades RNAs with a 5'P (processed RNAs) but not primary transcripts with a 5'PPP RNA [43].

Online Methods

Total RNA from S. pyogenes SF370 (M1 serotype) cells grown until mid-logarithmic phase was treated with DNase I to remove any residual genomic DNA [43].

The generated PCR fragments were cloned into pCR2.1-TOPO vector using TOPO TA cloning kit (Invitrogen) and the inserts of three to five clones were sequenced. (2) Primer extension: 5 to 10 µg of total RNA were denatured in presence of 5' radiolabelled reverse primer.

RNA was denatured 1 min at 95 °C and chilled on ice for 5 min, upon which 1 µg yeast RNA and 10 × structure buffer (0.1 M Tris pH 7, 1 M KCl, 0.1 M $MgCl_2$, Ambion) were added.

RNase T1 ladders were obtained by incubating labelled RNA (~0.2 pmol) in 1 × sequencing buffer (Ambion) for 1 min at 95 °C.

OH ladders were generated by 5 min incubation of 0.2 pmol labelled RNA in alkaline hydrolysis buffer (Ambion) at 95 °C.

Acknowledgements

A machine generated summary based on the work of Deltcheva, Elitza; Chylinski, Krzysztof; Sharma, Cynthia M.; Gonzales, Karine; Chao, Yanjie; Pirzada, Zaid A.; Eckert, Maria R.; Vogel, Jörg; Charpentier, Emmanuelle, 2011 in Nature.

A jumbo phage that forms a nucleus-like structure evades CRISPR–Cas DNA targeting but is vulnerable to type III RNA-based immunity.

https://doi.org/10.1038/s41564-019-0612-5

Abstract-Summary

We discovered a Serratia jumbo phage that evades type I CRISPR–Cas systems, but is sensitive to type III immunity.

All three native CRISPR–Cas complexes in Serratia—type I-E, I-F and III-A— were spatially excluded from the phage nucleus and phage DNA was not targeted.

Type III, but not type I, systems frequently targeted nucleus-forming jumbo phages that were identified in global viral sequence datasets.

The ability to recognize jumbo phage RNA and elicit immunity probably contributes to the presence of both RNA- and DNA-targeting CRISPR–Cas systems in many bacteria [44, 45].

Our results support the model that jumbo phage nucleus-like compartments serve as a barrier to DNA-targeting, but not RNA-targeting, defences, and that this phenomenon is widespread among jumbo phages.

Main

This is consistent with the Serratia jumbo phage being insensitive to type I systems (target DNA), while remaining sensitive to type III (targets both RNA and DNA) [46, 47].

As predicted, a Cas10 HD mutation (cas10$^{H17A, N18A}$) did not affect type III-A immunity, indicating that DNA cleavage by Cas10 is not necessary for jumbo phage resistance.

We have shown that jumbo phage immunity requires the specific RNA-targeting and cyclic oligonucleotide-signalling capabilities of the type III-A CRISPR–Cas system and requires the accessory nuclease.

We have discovered a jumbo phage that evades DNA targeting by two native type I CRISPR–Cas systems while retaining sensitivity to the RNA-targeting capabilities of the type III-A system.

Despite DNA protection, RNA export to the cytoplasm is a vulnerability of jumbo phages that can be exploited by type III CRISPR–Cas systems.

Jumbo phage infection has probably played a role in selecting for the observed widespread type III RNA-targeting immunity in strains already possessing DNA-based defences.

Methods

Enrichment for phages infecting Serratia was performed by mixing 100 μl of sewage sample with 5 ml of Serratia culture and incubating the mixture overnight at 30 °C, shaking at 160 r.p.m.

100 μl of overnight bacterial culture was mixed with serial dilutions of phage lysate and added to 4 ml of LBA overlay (0.35% w/v), which was then poured onto LBA plates.

To generate the cas10 knockout vector (pPF927), the primer pairs PF1934 (SalI site)/PF1935 (BamHI site) and PF1936 (BamHI site)/PF1937 (SphI) were used to amplify by PCR the cas10 upstream and downstream regions from Serratia wild-type colonies as the DNA template.

The two inserts were cloned with a three-part ligation including the suicide vector pPF923 (ref [48]) previously digested with SalI and SphI. The kanamycin^R-marked deletion strains were generated (PCF682; cas7 and PCF685; nuclease) using the plasmids pPF1929 and pPF1932 via homologous recombination as described for the tagged Cas complex strains.

Acknowledgements

A machine generated summary based on the work of Malone, Lucia M.; Warring, Suzanne L.; Jackson, Simon A.; Warnecke, Carolin; Gardner, Paul P.; Gumy, Laura F.; Fineran, Peter C.

2019 in Nature Microbiology.

A bacteriophage nucleus-like compartment shields DNA from CRISPR nucleases.

https://doi.org/10.1038/s41586-019-1786-y

Abstract-Summary

The viruses that infect bacteria, bacteriophages (phages), must avoid immune pathways that target nucleic acids, such as CRISPR–Cas and restriction-modification systems, to replicate efficiently [49].

We show that jumbo phage ΦKZ segregates its DNA from immunity nucleases of its host, Pseudomonas aeruginosa, by constructing a proteinaceous nucleus-like compartment.

ΦKZ is resistant to many immunity mechanisms that target DNA in vivo, including two subtypes of CRISPR–Cas3, Cas9, Cas12a and the restriction enzymes HsdRMS and EcoRI.

Cas proteins and restriction enzymes are unable to access the phage DNA throughout the infection, but engineering the relocalization of EcoRI inside the compartment enables targeting of the phage and protection of host cells.

Main

Phages that infect Pseudomonas aeruginosa can avoid destruction mediated by CRISPR by encoding 'anti-CRISPR' proteins that inhibit the type I-E and I-F CRISPR–Cas systems [50–52].

To determine whether any P. aeruginosa phages are resistant to the P. aeruginosa type I-C CRISPR–Cas system [53] (a common, but understudied variant [54]), we engineered a strain of P. aeruginosa to express type I-C cas3, cas5, cas8 and cas7 [55], and provided this strain with a panel of CRISPR RNAs (crRNAs) that target phages from five taxonomic groups: JBD30, D3, JBD68 (distinct temperate siphophages), F8 and ΦKZ (distinct lytic myophages).

Phage ΦKZ resists CRISPR–Cas targeting

Given the ability of this phage to evade unrelated CRISPR systems (types I and II)–including one from a microorganism that this phage does not infect (S. pyogenes)—we hypothesized that ΦKZ might be generally resistant to CRISPR–Cas immunity, as opposed to relying on specific inhibitor proteins.

The ability of ΦKZ to resist both CRISPR systems that are found in its natural host, Pseudomonas (types I-C and I-F), and those that are not naturally present in Pseudomonas (types II-A and V-A) suggests that this phage has a mechanism that enables 'pan-CRISPR' resistance.

We next tested type I and type II R-M systems (HsdRMS from P. aeruginosa and EcoRI from Escherichia coli, respectively; 24 and 92 cut sites in the ΦKZ genome, respectively).

Protein barrier occludes immune enzymes

To confirm that the ΦKZ phage genome could be a substrate for DNA cleavage, if accessed, two enzymes that did not cleave ΦKZ in vivo, EcoRI and Cas9, were assayed in vitro.

The ability of Cas9 to cleave ΦKZ genomic DNA (gDNA) in vitro was next assessed.

Owing to the large size of the ΦKZ genome (280 kb), we first subjected purified phage DNA to the restriction enzyme KasI to yield a 6.98-kb product, and then cleaved that species with a SpyCas9 nuclease loaded with both crRNA and trans-activating crRNA (tracrRNA) in vitro.

These results demonstrate that immune enzymes are capable of cleaving ΦKZ gDNA when they can access it, and that immune evasion is probably not a result of an intrinsic feature of the phage DNA (such as base modifications that can impede the Cascade–Cas3 complex, Cas9 and EcoRI) [56–59].

Phage targeting via enzyme localization

We fused the single effector enzyme Cas9 to ORF152, a phage-encoded RecA-like protein that is internalized within the shell [60, 61].

Phage mRNA is sensitive to Cas13a.

The nucleus-like structure produced by ΦKZ provides robust resistance to DNA-targeting immune pathways, but other immune systems may exist in vivo that can evade this mechanism.

To test this, we adapted the type VI-A CRISPR RNA-guided RNA nuclease Cas13a from Listeria seeligeri (LseCas13a) [62, 63] for phage targeting in P. aeruginosa.

During Cas13a targeting of the shell mRNA, imaging revealed that infections arrested before the phage DNA proceeded from its injection site at the poles.

The absence of diffusion or clearance of the phage DNA (for example, by the endogenous type I R-M system) suggests that the injected phage genome may be protected before shell assembly by injected phage proteins or pre-existing host factors.

Discussion

Given the efficacy of the RNA-targeting CRISPR–Cas13 system, we propose that type VI CRISPR systems are well suited to target the mRNA of DNA phages when the DNA is inaccessible (that is, owing to base modifications or physical segregation).

The polar localization of the injected phage DNA during mRNA targeting suggests a poorly understood early protective mechanism.

Regardless, we conclude that the phage-assembled nucleus-like structure provides a strong protective barrier to DNA-targeting immune pathways.

Methods

The resulting strains were assayed for phage sensitivity under standard phage plating conditions, with induction of both Cas13 and the gRNAs (50 μg ml^{-1} gentamicin, 0.1% (l)-arabinose, 1 mM IPTG).

Plasmids expressing Cherry alone, Cherry-Cas3, or Cherry-Cas8 (of type I-C and type I-F systems) and Cherry-TopA were constructed by Gibson assembly in the pHERD30T plasmid digested with SacI and PstI. These fusions have a GGAG-GCGGTGGAGCC nucleotide (G-G-G-G-A amino acid) linker sequence in between them.

After incubating at 37 °C for 10 min, samples were mixed with 4 ml of 0.7% agar, 10 mM $MgSO_4$, 1 mM IPTG and 0.1% arabinose and plated in LB agar plates with gentamicin (50 $\mu g\ ml^{-1}$).

Overnight cultures (5 ml) of a strain expressing Cas9 and an sgRNA targeting ΦKZ (SDM065), and a strain expressing cMyc–ORF152 (bESN27), were grown at 30 °C in LB medium with gentamicin.

Acknowledgements

A machine generated summary based on the work of Mendoza, Senén D.; Niewe-glowska, Eliza S.; Govindarajan, Sutharsan; Leon, Lina M.; Berry, Joel D.; Tiwari, Anika; Chaikeeratisak, Vorrapon; Pogliano, Joe; Agard, David A.; Bondy-Denomy, Joseph, 2019 in Nature.

An anti-CRISPR viral ring nuclease subverts type III CRISPR immunity.

https://doi.org/10.1038/s41586-019-1909-5

Abstract-Summary

Type III CRISPR systems detect viral RNA, resulting in the activation of two regions of the Cas10 protein: an HD nuclease domain (which degrades viral DNA) [46, 64] and a cyclase domain (which synthesizes cyclic oligoadenylates from ATP) [65–67].

Cyclic oligoadenylates in turn activate defence enzymes with a CRISPR-associated Rossmann fold domain [68], sculpting a powerful antiviral response [69–72] that can drive viruses to extinction [69, 70].

We identify a new family of viral anti-CRISPR (Acr) enzymes that rapidly degrade cyclic tetra-adenylate (cA_4).

The enzyme uses a previously unknown fold to bind cA_4 specifically, and a conserved active site to rapidly cleave this signalling molecule, allowing viruses to neutralize the type III CRISPR defence system.

The AcrIII-1 family has a broad host range, as it targets cA_4 signalling molecules rather than specific CRISPR effector proteins.

Extended:

The experiments were not randomized and the investigators were not blinded to allocation during experiments and outcome assessment.

Main

We identified in the archaeon Sulfolobus solfataricus a family of cellular enzymes—referred to hereafter as the CRISPR-associated ring nuclease 1 (Crn1) family—that degrades cA_4 molecules and deactivates the cA_4-dependent RNase Csx1 [73].

Viruses have responded to the threat of the CRISPR system by evolving a range of anti-CRISPR (Acr) proteins, which are used to inhibit and overcome the cell's CRISPR defences using a variety of mechanisms (reviewed in ref [74]).

Acrs have been identified for many of the CRISPR effector subtypes, and number more than 40 families [75].

DUF1874 is a type III anti-CRISPR, AcrIII-1

These data are consistent with the hypothesis that SIRV1 gp29 functions as an Acr specific for the type III CRISPR defence.

The presence of the duf1874 gene on the plasmid reduced immunity for cA_4-mediated, but not cA_6-mediated, CRISPR defence.

This observation supports the hypothesis that DUF1874 acts as an Acr against cA_4-mediated type III CRISPR defence.

AcrIII-1 degrades cA_4 rapidly

By varying the target RNA input and following cA_4 levels and Csx1 activity, we compared the abilities of Crn1 and AcrIII-1 to destroy the signalling molecule and deactivate the ancillary defence nuclease Csx1.

AcrIII-1 degraded cA_4 completely at the highest target RNA concentration examined, preventing Csx1 activation.

Structure and mechanism of AcrIII-1

The structure of AcrIII-1 is unrelated to that of proteins with the CRISPR-associated Rossmann fold (CARF) domain—the only protein family known thus far to bind cOA [68].

Comparison of the cA_4-bound and apo structures reveals a substantial movement of a loop (comprising residues 82–94) and subsequent α-helix to bury cA_4 within the dimer.

We also noted that the conserved residue E88, situated on the tip of the loop that covers the binding site, is positioned close to the H47 residue of the opposite subunit.

By targeting a key signalling molecule, a single AcrIII-1 enzyme should have broad utility in the inhibition of endogenous cA_4-specific type III CRISPR systems in any species.

A type III Acr (AcrIIIB1) has been reported that appears to function by binding and inhibiting the type III-B effector complex [76].

Phylogenetic analysis of AcrIII-1

The acrIII-1 gene is clearly part of an integrated mobile genetic element, such as the yddF gene in B. subtilis [77].

Both active sites are conserved, this fusion protein may have cA_4-activated RNase activity coupled with a cA_4-degradative ring nuclease, thus providing an explicit linkage between the AcrIII-1 family and the type III CRISPR system.

Cyclic nucleotides in defence systems

AcrIII-1 is, to our knowledge, the first Acr to be predicted to have functional roles in both 'offense and defence'.

It remains to be determined whether the acrIII-1 gene arose in viruses and was appropriated by cellular type III systems, or vice versa.

The unprecedentedly wide occurrence of this Acr across many archaeal and bacterial virus families reflects the fact that this enzyme degrades a key signalling molecule to subvert cellular immunity.

Recent discoveries have highlighted the existence of diverse cellular defence systems involving cyclic nucleotide signalling in bacteria [78–80].

Methods

For single-turnover kinetics experiments, we assayed AcrIII-1 SIRV1 gp29 and variants (4 μM protein dimer) for radiolabelled cA_4 degradation by incubating with 1/400 diluted ^{32}P-labelled SsoCsm cOA (roughly 200 nM cA_4, generated in a 100 μl cOA-synthesis reaction as above) in Csx1 buffer supplemented with 1 mM EDTA at 50 °C.

In the absence or presence of Crn1 Sso2081 (2 μM dimer) or AcrIII-1 SIRV1 gp29 (2 μM dimer), we incubated 4 μg S. solfataricus Csm complex (roughly 140 nM Csm carrying crRNA targeting A26 RNA target) with A26 RNA target (50 nM, 20 nM, 5 nM, 2 nM or 0.5 nM) in buffer containing 20 mM MES pH 6.0, 100 mM NaCl, 1 mM DTT and three units SUPERase•In Inhibitor supplemented with 2 mM $MgCl_2$ and 0.5 mM ATP at 70 °C for 60 min.

Acknowledgements

A machine generated summary based on the work of Athukoralage, Januka S.; McMahon, Stephen A.; Zhang, Changyi; Grüschow, Sabine; Graham, Shirley; Krupovic, Mart; Whitaker, Rachel J.; Gloster, Tracey M.; White, Malcolm F, 2020 in Nature.

2.2 Gene Editing and Beyond Editing

Machine generated keywords: zygote, gene expression, lineage, sgrna, singlecell, transcript, brain, gene edit, neuron, mouse, cpf, mrna, element, drive, cleavage.

DNA interrogation by the CRISPR RNA-guided endonuclease Cas9.

https://doi.org/10.1038/nature13011

Abstract-Summary

The clustered regularly interspaced short palindromic repeats (CRISPR)-associated enzyme Cas9 is an RNA-guided endonuclease that uses RNA–DNA base-pairing to target foreign DNA in bacteria.

We show that both binding and cleavage of DNA by Cas9–RNA require recognition of a short trinucleotide protospacer adjacent motif (PAM).

Non-target DNA binding affinity scales with PAM density, and sequences fully complementary to the guide RNA but lacking a nearby PAM are ignored by Cas9–RNA.

These results reveal how Cas9 uses PAM recognition to quickly identify potential target sites while scanning large DNA molecules, and to regulate scission of double-stranded DNA.

Main

RNA-mediated adaptive immune systems in bacteria and archaea rely on CRISPRs and CRISPR-associated (Cas) proteins to provide protection from invading viruses and plasmids [81].

Transcription of the CRISPR array followed by enzymatic processing yields short CRISPR RNAs (crRNAs) that direct Cas protein-mediated cleavage of complementary target sequences within invading viral or plasmid DNA [28, 42, 82].

In type II CRISPR-Cas systems, Cas9 functions as an RNA-guided endonuclease that uses a dual-guide RNA consisting of crRNA and trans-activating crRNA (tracrRNA) for target recognition and cleavage by a mechanism involving two nuclease active sites that together generate double-stranded DNA (dsDNA) breaks [83, 84].

Single-molecule visualization of Cas9

DNA targeting by Cas9–RNA is faithfully recapitulated in the DNA curtains assay.

We next used apo-Cas9 protein to confirm that the binding observed in DNA curtains assays was due to Cas9–RNA and not apo-Cas9 lacking guide RNA.

To test whether DNA-bound apo-Cas9 could be distinguished from Cas9–RNA, we measured the lifetime of apo-Cas9 on DNA curtains before and after injection of crRNA–tracrRNA or heparin.

Cas9–RNA finds targets by three-dimensional diffusion

Site-specific DNA-binding proteins can locate target sites by three-dimensional collisions or through facilitated diffusion processes including one-dimensional sliding, hopping, and/or intersegmental transfer [85]; these mechanisms can be distinguished by single-molecule imaging [86–88].

The number of observed binding events was not uniformly distributed along the substrate, indicating that some underlying feature of the λ-DNA might influence the target search.

These results, together with the insensitivity of short-lived binding events to ionic strength, suggested that Cas9–RNA might bind specifically to PAMs and minimize interactions with non-PAM DNA while searching for potential targets.

A PAM is required for DNA interrogation

These results demonstrate that the residence time of Cas9–RNA on non-target DNA lacking PAMs is negligible, and support the hypothesis that transient, non-target DNA binding events observed on the DNA curtains probably occurred at PAM sequences.

These results demonstrate that PAM recognition is an obligate first step during target recognition by Cas9–RNA, as previously proposed [83].

Mechanism of RNA–DNA heteroduplex formation

After PAM recognition, Cas9–RNA must destabilize the adjacent duplex and initiate strand separation to enable base pairing between the target DNA strand and the crRNA guide sequence.

When mismatches to the crRNA are encountered within the first two nucleotides of the target sequence, Cas9–RNA loses the ability to interrogate and recognize the remainder of the DNA.

The pattern of inhibition observed with the different competitor DNAs indicates that sequence homology adjacent to the PAM is necessary to initiate target duplex unwinding until the reaction has proceeded sufficiently far ($\sim$12 bp, approximately one turn of an A-form RNA–DNA helix), such that the energy necessary for further propagation of the RNA–DNA heteroduplex falls below the energy needed for the reverse reaction.

The PAM triggers Cas9 nuclease activity

One might expect PAM recognition to be dispensable for Cas9–RNA-mediated recognition and cleavage of a single-stranded DNA (ssDNA) target.

Cas9–RNA recognizes the 5′-NGG-3′ PAM on the non-target DNA strand [83], so the ssDNA substrates did not contain a PAM but rather the complement to the PAM sequence.

Discussion

Rather than sampling all DNA equivalently, Cas9–RNA accelerates the search by rapidly dissociating from non-PAM sites, thereby reducing the amount of time spent at off-targets.

Only upon binding to a PAM site does Cas9–RNA interrogate the flanking DNA for guide RNA complementarity, as was previously proposed for Cas9 (ref 86) and a distinct CRISPR RNA-guided complex (Cascade) [89].

The complex dissociation kinetics observed on non-target λ-DNA would arise from heterogeneity in the potential target sites as Cas9–RNA probes sequences adjacent to PAMs for guide RNA complementarity.

Our data indicate that efforts to minimize off-target effects during genome engineering using Cas9–RNA complexes need only consider off-targets adjacent to a PAM, because potential targets lacking a PAM are unlikely to be interrogated [90–93].

Methods Summary

Cas9–RNA complexes for single-molecule experiments were reconstituted by incubating Cas9 and a 10 × molar excess of crRNA–tracrRNA in reaction buffer (20 mM Tris–HCl pH 7.5, 100 mM KCl, 5 mM $MgCl_2$, 5% glycerol, 1 mM dithiothreitol (DTT)) for $\sim$10 min at 37 °C, before injecting 1–2 nM into the flow cell.

Double-tethered DNA curtains were prepared as described [87, 88], and position and lifetime measurements were determined from kymographs generated for each DNA molecule.

Bulk competition cleavage assays were conducted at room temperature in reaction buffer and contained $\sim$1 nM radiolabelled λ1 target DNA, 10 nM Cas9–RNA complex and 500 nM competitor DNA.

Competition experiments were analysed to determine the survival probability of the target DNA.

Online Methods

100 nM 3 $\times$ -Flag-tagged dCas9 was reconstituted with 1 μM crRNA–tracrRNA targeting the desired region of λ-DNA by incubating for $\sim$10 min at 37 °C in reaction buffer (20 mM Tris–HCl pH 7.5, 100 mM KCl, 5 mM MgCl$_2$, 5% glycerol, 1 mM dithiothreitol (DTT)).

10 nM dCas9–RNA was then incubated with λ-DNA (100 pM) for $\sim$15 min at 37 °C in 40 mM Tris–HCl pH 7.5, 25 mM KCl, 1 mg ml^{-1} BSA, 1 mM MgCl$_2$ and 1 mM DTT, before being diluted to 1 nM and injected into the flow cell.

Binding reactions contained 0.1–1 nM DNA and increasing apo-dCas9 or dCas9–RNA concentrations, and were incubated at 37 °C for 1 h before being resolved by 5% native polyacrylamide gel electrophoresis (0.5 $\times$ TBE buffer with 5 mM MgCl$_2$) run at 4 °C.

Acknowledgements

A machine generated summary based on the work of Sternberg, Samuel H.; Redding, Sy; Jinek, Martin; Greene, Eric C.; Doudna, Jennifer A, 2014 in Nature.

In vivo *interrogation of gene function in the mammalian brain using CRISPR-Cas9.*

https://doi.org/10.1038/nbt.3055

Main

To assess the editing efficiency of our dual-vector system, we transduced mouse primary cortical neurons with SpCas9 and Mecp2-targeting sgRNA or control sgRNA (targeting the bacterial lacZ gene).

These results show that the combination of SpCas9-mediated genome perturbation, intact nuclei purification and RNA-seq analysis provides a robust method to study transcriptional regulation in adult neurons in vivo and identify candidate genes that might modulate specific neuronal functions or disease processes.

Cas9-mediated MeCP2 depletion in adult mice underscores the maintenance role of MeCP2 in adult brain [94] and reveals the role of MeCP2 in the integration of inputs that underlie a cortical response feature; it further demonstrates the versatility of SpCas9 in facilitating targeted gene knockdown in the mammalian brain in vivo for studying gene function in health and disease.

To test the efficacy of multiplex genome editing in vivo, we stereotactically injected a mixture of AAV-SpCas9 and AAV-SpGuide (targeting Dnmt3a, Dnmt1 and Dnmt3b) into the DG of adult mice.

Methods

1 ml ddH$_2$O was added and lysate was kept on ice for 5 min before cell lysates were homogenized with Dounce homogenizer (Sigma); 20 times with pestle A, followed by 10 times with pestle B. 2 ml equilibration buffer (120 mM β-glycerophosphate pH 7.0, 2 mM MgCl$_2$, 1 mM PMSF, 1 mM β-mercaptoethanol, 50% glycerol) was added and nuclei were centrifuged (1,000 g for 10 min at 4 °C) using a sucrose gradient (lower: 500 mM sucrose, 2 mM MgCl$_2$, 25 mM KCl, 65 mM β-glycerophosphate pH 7.0, 20% glycerol, 1 mM PMSF, 1 mM β-mercaptoethanol; upper: 340 mM sucrose, 2 mM MgCl$_2$, 25 mM KCl, 65 mM β-glycerophosphate (pH 7.0), 20% glycerol, 1 mM PMSF, 1 mM β-mercaptoethanol).

For cell nuclei purification, dissected tissue was gently homogenized in 2 ml ice-cold homogenization buffer (HB) (320 mM sucrose, 5 mM CaCl, 3 mM Mg(Ac)$_2$, 10 mM Tris pH7.8, 0.1 mM EDTA, 0.1% NP40, 0.1 mM PMSF, 1 mM β-mercaptoethanol) using 2 ml Dounce homogenizer (Sigma); 25 times with pestle A, followed by 25 times with pestle B. Next, 3 ml of HB was added up to 5 ml total and kept on ice for 5 min.

Acknowledgements

A machine generated summary based on the work of Swiech, Lukasz; Heidenreich, Matthias; Banerjee, Abhishek; Habib, Naomi; Li, Yinqing; Trombetta, John; Sur, Mriganka; Zhang, Feng, 2014 in Nature Biotechnology.

Genome-scale transcriptional activation by an engineered CRISPR-Cas9 complex.

https://doi.org/10.1038/nature14136

Abstract-Summary

We describe structure-guided engineering of a CRISPR-Cas9 complex to mediate efficient transcriptional activation at endogenous genomic loci.

We used these engineered Cas9 activation complexes to investigate single-guide RNA (sgRNA) targeting rules for effective transcriptional activation, to demonstrate multiplexed activation of ten genes simultaneously, and to upregulate long intergenic non-coding RNA (lincRNA) transcripts.

We also synthesized a library consisting of 70,290 guides targeting all human RefSeq coding isoforms to screen for genes that, upon activation, confer resistance to a BRAF inhibitor.

A gene expression signature based on the top screening hits correlated with markers of BRAF inhibitor resistance in cell lines and patient-derived samples.

Main

Genome-scale GOF screening approaches have largely remained limited to the use of cDNA library overexpression systems.

Novel technologies that overcome such limitations would enable systematic, genome-scale GOF perturbations at endogenous loci.

Among the established custom DNA-binding domains, Cas9 is most easily scaled to facilitate genome-scale perturbations [95, 96] owing to its simplicity of programming relative to zinc finger proteins and transcription activator-like effectors (TALEs).

Tiling a given promoter region with several sgRNAs can produce more robust transcriptional activation [97–99], but this requirement presents enormous challenges for scalability, and in particular for establishing pooled, genome-wide GOF screens.

Using this new activation system, we demonstrate activation of endogenous genes as well as non-coding RNAs, elucidate design rules for effective sgRNA target sites, and establish and apply genome-wide dCas9-based transcription activation screening to study drug resistance in a melanoma model.

Structure-guided design of Cas9 complex

Transformation of the Cas9–sgRNA complex into an effective transcriptional activator requires finding optimal anchoring positions for the activation domains.

We next tested whether MS2-mediated recruitment of VP64 to the tetraloop and stem loop 2 could mediate transcriptional upregulation more efficiently than a dCas9–VP64 fusion.

Recruitment of VP64 to both positions (sgRNA 2.0) resulted in an additive effect, leading to a 12-fold increase over dCas9–VP64 (sgRNA 1.0).

We chose the NF-κB trans-activating subunit p65 that, while sharing some common co-factors with VP64, recruits a distinct subset of transcription factors and chromatin remodelling complexes.

On the basis of these collective results, we concluded that the combination of sgRNA 2.0, NLS–dCas9–VP64 and MS2–p65–HSF1 comprises the most effective transcription activation system, and designated it synergistic activation mediator (SAM).

Design rules for efficient sgRNAs

To evaluate thoroughly the effectiveness of SAM for activating endogenous gene transcription, we chose 12 genes that were previously found by several groups to be difficult to activate using dCas9–VP64 and individual sgRNA 1.0 guides [98–100].

SAM performed consistently better than sgRNA 1.0 + dCas9–VP64 for all 96 guides, with a median gain of 105-fold greater upregulation across all 12 genes (activation by SAM divided by activation by sgRNA 1.0 + dCas9–VP64).

Previous studies have demonstrated that the poor activation efficiency of single sgRNAs can be overcome by combining dCas9–VP64 with a pool of sgRNAs tiling the proximal promoter region of the target gene [97–99].

We compared the single sgRNA activation efficiency of SAM against dCas9–VP64 combined with a pool of 8 sgRNA 1.0 guides, all targeting the same gene.

For inter-gene variability, differences in activation magnitudes could be due to epigenetic factors and/or variation in basal transcription levels.

Multiplex gene activation

Most genes (excluding IL1R2) exhibited a decrease in the amount of upregulation achieved when concurrently targeted with 9 other genes.

Specificity of SAM-mediated activation

To assess SAM specificity, we chose HBG1/2 as our target gene, reasoning that globin genes would have few downstream targets that could confound our specificity analysis.

These results suggest that SAM-mediated gene activation is specific with minimal off-target activity.

Genome-scale gene activation screen

We applied genome-scale CRISPR knockout (GeCKO) screening [95] in A375 (BRAF(V600E)) melanoma cells to identify LOF mutations capable of mediating resistance against the BRAF inhibitor PLX-4720.

EGFR has been previously validated as a mediator of resistance to PLX-4720 through PI3K-AKT, in addition to ERK [101, 102].

Four out of the top 10 hits from our screen belong to the family of G-protein-coupled receptors (GPCRs: GPR35, LPAR1, LPAR5 and P2RY8), which emerged as the top-ranked protein class conferring resistance to multiple MAP kinase inhibitors in melanoma cells in a recent screen using cDNA overexpression [103].

To verify the results from the PLX-4720 resistance screen, we validated each of the top 13 genes.

Discussion

We have taken a structure-guided approach to design a dCas9-based transcription activation system for achieving robust, single sgRNA-mediated gene upregulation.

By engineering the sgRNA to incorporate protein-interacting aptamers, we assembled a synthetic transcription activation complex consisting of multiple distinct effector domains modelled after natural transcription activation processes.

Combining wild-type Cas9-mediated genome modifications with SAM-mediated recruitment of epigenetic modifiers will constitute powerful approaches for studying genome organization and regulation in diverse biological processes.

Methods

In the CCLE data set [104], gene expression data (RNA-sequencing, GCHub: https://cghub.ucsc.edu/datasets/ccle.html) and pharmacological data (activity area for MAPK pathway inhibitors) from BRAF(V600) mutant melanoma cell lines were used to compute the association between PLX-4720 resistance and the gene expression of each of the top hits.

Gene expression data (Affymetrix GeneChip HT-HGU133) and PLX-4720 pharmacological data (GI_{50}: half-maximal growth inhibition concentration; only for a subset of the samples) from short term melanoma cultures (STC) [105] were also used for plotting the gene expression of top hits and their ssGSEA signature scores.

Gene expression (RNA-sequencing) and genotyping data were collected from 113 BRAF(V600)-mutant primary and metastatic patient tumours from The Cancer Genome Atlas (https://tcga-data.nci.nih.gov/tcga/) and these data were similarly used for determining the association between resistance and the expression of top hits/ssGSEA signature scores.

Cells were selected for guide expression with Zeocin (Life Technologies) for 5 days and replated at low density (3×10^3 cells per well in a 96-well plate).

Acknowledgements

A machine generated summary based on the work of Konermann, Silvana; Brigham, Mark D.; Trevino, Alexandro E.; Joung, Julia; Abudayyeh, Omar O.; Barcena, Clea; Hsu, Patrick D.; Habib, Naomi; Gootenberg, Jonathan S.; Nishimasu, Hiroshi; Nureki, Osamu; Zhang, Feng, 2014 in Nature.

CRISPR/Cas9-mediated gene editing in human tripronuclear zygotes.

https://doi.org/10.1007/s13238-015-0153-5

Abstract-Summary

Genome editing tools such as the clustered regularly interspaced short palindromic repeat (CRISPR)-associated system (Cas) have been widely used to modify genes in model systems including animal zygotes and human cells, and hold tremendous promise for both basic research and clinical applications.

A serious knowledge gap remains in our understanding of DNA repair mechanisms in human early embryos, and in the efficiency and potential off-target effects of using technologies such as CRISPR/Cas9 in human pre-implantation embryos.

We used tripronuclear (3PN) zygotes to further investigate CRISPR/Cas9-mediated gene editing in human cells.

We found that CRISPR/Cas9 could effectively cleave the endogenous β-globin gene (HBB).

The efficiency of homologous recombination directed repair (HDR) of HBB was low and the edited embryos were mosaic.

Introduction

The ease, expedience, and efficiency of the CRISPR/Cas9 system have lent itself to a variety of applications, including genome editing, gene function investigation, and gene therapy in animals and human cells (Chang and others, 109; Cho and others, 110; Cong and others, 111; Friedland and others, 112; Hsu and others, 113; Hwang and others, 114; Ikmi and others, 115; Irion and others, 116; Jinek and others, 117; Li and others, 118; Li and others, 119; Long and others, 120; Ma and others, 121; Mali and others, 122; Niu and others, 123; Smith and others, 124; Wu and others, 125; Wu and others, 126; Yang and others, 127).

Three groups recently found through whole genome sequencing that off-target effects of CRISPR/Cas9 appeared rare in human pluripotent stem cells (Smith and others, 128; Suzuki and others, 129; Veres and others, 130), raising the possibility that high frequencies of unintended targeting by CRISPR/Cas9 may be more prevalent in cancer cell lines.

Despite great progress in understanding the utilization of CRISPR/Cas9 in a variety of model organisms, much remains to be learned regarding the efficiency and specificity of CRISPR/Cas9-mediated gene editing in human cells, especially in embryos.

We report that the CRISPR/Cas9 system can cleave endogenous gene efficiently in human tripronuclear zygotes, and that the DSBs generated by CRISPR/Cas9 cleavage are repaired by NHEJ and HDR.

Results

The HBD footprints left in the repaired HBB locus should enable us to investigate whether and how endogenous homologous sequences may be utilized as HDR templates, information that will prove invaluable to any future endeavors that may employ CRISPR/Cas9 to repair gene loci with repeated sequences.

CRISPR/Cas9 targeting of the β-globin locus was previously reported to have substantially high off-target activity in cultured human cells (Cradick and others, 131).

All GFP-positive embryos were then collected for whole-genome amplification by multiplex displacement amplification (Dean and others, 132; Hosono and others, 133), followed by PCR amplification of the G1 gRNA target region and sequencing.

To determine the off-target effects of CRISPR/Cas9 in these embryos, we again examined the top 7 potential off-target sites plus the site in the HBD gene.

This lack of bi-directional sequence exchange supports the notion that the HBB gene was repaired primarily through non-crossover HDR (San Filippo and others, 134).

Discussion

We used 3PN zygotes to investigate the specificity and fidelity of the CRISPR/Cas9 system.

Further investigation of the molecular mechanisms of CRISPR/Cas9-mediated gene editing in human model is sorely needed.

Off-target effect of CRISPR/Cas9 should be investigated thoroughly before any clinical application (Baltimore and others, 135; Cyranoski, 136; Lanphier and others, 137).

Materials and methods

The pDR274 vector encoding gRNA sequences was in vitro transcribed using the MEGAshortscript T7 kit (Life Technologies).

To avoid false positive calls that overlap with repeat sequences and/or include homopolymers (Bansal and Libiger, 138), we removed indels and SNVs that overlapped with low-complexity regions as defined by RepeatMasker (UCSC Genome Browser) and filtered out indels and SNVs containing homopolymers (>7 bp) in the low-complexity flanking region ($\pm$100 bp), removing 55.58% of potential indels and 17.01% of potential SNVs.

Of the 12 candidate indels identified by this analysis, there were ten on-target indels in all samples and two off-target indels in samples A and C. Candidate off-target sites were further confirmed by PCR and sequencing.

Acknowledgements

A machine generated summary based on the work of Liang, Puping; Xu, Yanwen; Zhang, Xiya; Ding, Chenhui; Huang, Rui; Zhang, Zhen; Lv, Jie; Xie, Xiaowei; Chen, Yuxi; Li, Yujing; Sun, Ying; Bai, Yaofu; Songyang, Zhou; Ma, Wenbin; Zhou, Canquan; Huang, Junjiu, 2015 in Protein & Cell.

A CRISPR-Cas9 gene drive system targeting female reproduction in the malaria mosquito vector Anopheles gambiae.

https://doi.org/10.1038/nbt.3439

Main

Each drive construct was designed to home, in both sexes, into the cognate WT locus and contained the following components: (i) the Cas9 nuclease gene under the control of the vasa2 promoter, shown in a previous report to be active in the germline of both sexes [136]; (ii) a gRNA sequence designed to direct the cleavage activity of the nuclease to the same sequence targeted in the gene-knockout experiments and under the promoter of the ubiquitously expressed, PolIII-transcribed U6 gene [137]; and (iii) a visual marker (3xP3::RFP).

The rates of super-Mendelian inheritance that we observed with CRISPR-based homing constructs at female-fertility loci establish a solid basis for the development of a gene drive system that has the potential to substantially reduce mosquito populations.

The success of gene drive technology for vector control will depend on the choice of suitable promoters to effectively drive homing during the process of gametogenesis, the phenotype of the disrupted genes, the robustness of the nuclease during homing and the ability of the target population to generate compensatory mutations.

Methods

Gene-targeting vectors were assembled by Gateway cloning (Invitrogen) and designed to contain an attP-flanked 3xP3::GFP marker construct enclosed within homology arms extending 2 kb either direction of the expected CRISPR[h] cleavage site, as well as an external 3xP3::RFP marker.

For the generation of the hdrGFP docking lines the donor construct (300 ng/μl) containing regions of homology to the relevant target locus was injected together with the relevant CRISPR plasmid (300 ng/μμl) for AGAP007280 (p16501) and AGAP005958 (p16505) or, for AGAP011377, plasmids expressing the left and right monomers of the TALEN (each at 300 ng/μl).

Successful cassette exchange of CRISPR[h] alleles was interrogated by PCR using primers binding the CRISPR[h] construct, RFP2q-F (GTGCTGAAGGGCGAGATC-CACA) and hCas9-F7 (CGGCGAACTGCAGAAGGGAA) with primers binding the genome: AGAP011377 using Seq-5958-F and Seq-5958-R, AGAP005958 using Seq-5958-F and Seq-5958-R, and AGAP007280 using Seq-7280-F and Seq-7280-R. To assess molecularly the activity of CRISPR at the target locus, the target site was sequenced in those progeny (RFP⁻) that apparently failed to receive a CRISPR homing allele from a hemizygous RFP⁺ parent.

Acknowledgements

A machine generated summary based on the work of Hammond, Andrew; Galizi, Roberto; Kyrou, Kyros; Simoni, Alekos; Siniscalchi, Carla; Katsanos, Dimitris; Gribble, Matthew; Baker, Dean; Marois, Eric; Russell, Steven; Burt, Austin; Windbichler, Nikolai; Crisanti, Andrea; Nolan, Tony, 2016 in Nature Biotechnology.

The CRISPR-associated DNA-cleaving enzyme Cpf1 also processes precursor CRISPR RNA.

https://doi.org/10.1038/nature17945

Main

Mutagenesis of all three nucleotides followed by DNA cleavage analysis shows that Cpf1 recognizes a PAM, defined as 5′-YTN-3′, upstream of the crRNA-complementary DNA sequence on the non-target strand.

If there were two active sites in Cpf1, each coordinating one of the metal ions and cleaving one of the DNA strands, we would expect a difference in cleavage of target and non-target strands depending on the ion used.

Cpf1 is the first enzyme with two specificities, cleaving RNA in a sequence- and structure-dependent manner, and also performing DNA cleavage in the presence of the RNA that is produced in the first reaction.

In the context of CRISPR immunity, type V-A appears to be the most minimalistic system described thus far, using only one enzyme, Cpf1, to process pre-crRNA and then using this RNA to specifically target and cut invading DNA.

Methods

Plasmid DNA cleavage assays were performed by pre-incubating 100 nM Cpf1 with 200 nM RNA in KGB buffer supplemented with either 10 mM $MgCl_2$ or 10 mM $CaCl_2$ for 15 min at 37 °C.

In cleavage assays using radioactively labelled substrates, 5 nM of 5′-labelled double-stranded oligonucleotides were added to the pre-formed complex of Cpf1 and RNA, and incubated at 37 °C for 1 h. After proteinase K treatment, 10 μl of 2 × denaturing loading buffer (95% formamide, 0.025% SDS, 0.5 mM EDTA, 0.025% bromophenol blue) were added.

A total of 0.5 nM radiolabelled RNA were incubated with Cpf1 in binding buffer (20 mM Tris (pH 7.5), 150 mM KCl, 10 mM $CaCl_2$, 1 mM DTT, 5% glycerol, 0.01% Triton X-100, 10 μg ml⁻¹ BSA) for 1 h at 37 °C and loaded on 4% native polyacrylamide gels running at 10 V cm⁻¹ for 30 min in 0.5 × TBE.

Acknowledgements

A machine generated summary based on the work of Fonfara, Ines; Richter, Hagen; Bratovič, Majda; Le Rhun, Anaïs; Charpentier, Emmanuelle, 2016 in Nature.

Multiplex gene editing by CRISPR–Cpf1 using a single crRNA array.

https://doi.org/10.1038/nbt.3737

Main

We leveraged the simplicity of Cpf1 crRNA maturation to achieve multiplex genome editing in HEK293T cells using customized CRISPR arrays.

The crRNA targeting EMX1 resulted in indel frequencies of < 2% when expressed from array 3.

We next tested multiplex genome editing in neurons using AsCpf1.

We tested whether AsCpf1 could be expressed in the brains of living mice for multiplex genome editing in vivo.

Our results demonstrate the effectiveness of AAV-mediated delivery of AsCpf1 into the mammalian brain and simultaneous multi-gene targeting in vivo using a single array transcript.

These data highlight the utility of Cpf1 array processing in designing simplified systems for in vivo multiplex gene editing.

This system should simplify guide RNA delivery for many genome editing applications in which targeting of multiple genes is desirable.

Methods

Cells were incubated at 37 °C for 72 h after transfection before genomic DNA extraction.

For inhibition of glia cell proliferation, cytosine-beta-D-arabinofuranoside (AraC, Sigma) at a final concentration of 10 μM was added to the culture medium after 48 h and replaced by fresh culture medium after 72 h. For AAV1 transduction, cultured neurons were infected with low-titer AAV1 as described previously [100].

AAV-injected dentate gyrus tissues were lysed in 100 μl of ice-cold RIPA buffer (Cell Signaling Technologies) containing 0.1% SDS and protease inhibitors (Roche, Sigma) and sonicated in a Bioruptor sonicator (Diagenode) for 1 min.

Protein samples were separated under reducing conditions on 4–15% Tris–HCl gels (Bio-Rad) and analyzed by western blotting using primary antibodies: mouse anti-HA (Cell Signaling Technologies 1:500), mouse anti-GFP (Roche, 1:500), rabbit anti-Tubulin (Cell Signaling Technologies, 1:10,000) followed by secondary anti-mouse and anti-rabbit HRP antibodies (Sigma-Aldrich, 1:10,000).

Acknowledgements

A machine generated summary based on the work of Zetsche, Bernd; Heidenreich, Matthias; Mohanraju, Prarthana; Fedorova, Iana; Kneppers, Jeroen; DeGennaro, Ellen M; Winblad, Nerges; Choudhury, Sourav R; Abudayyeh, Omar O; Gootenberg,

Jonathan S; Wu, Wen Y; Scott, David A; Severinov, Konstantin; van der Oost, John; Zhang, Feng, 2016 in Nature Biotechnology.

CRISPR/Cas9-mediated gene editing in human zygotes using Cas9 protein.

https://doi.org/10.1007/s00438-017-1299-z

Abstract-Summary

Previous works using human tripronuclear zygotes suggested that the clustered regularly interspaced short palindromic repeat (CRISPR)/Cas9 system could be a tool in correcting disease-causing mutations.

We demonstrate that CRISPR/Cas9 is also effective as a gene-editing tool in human 2PN zygotes.

By injection of Cas9 protein complexed with the appropriate sgRNAs and homology donors into one-cell human embryos, we demonstrated efficient homologous recombination-mediated correction of point mutations in HBB and G6PD.

Introduction

Correcting disease-causing genetic defects in human zygotes was previously unthinkable because the efficiency would be too low to be of any practical value.

Separately from the debate of the merit and ethics of germline editing in humans, the feasibility of correcting genetic defects via CRISPR technology in human zygotes has not been really tested.

Two recent reports used 3PN (3 or more pronuclei) human zygotes to test gene editing efficiency and the reported HDR efficiencies were only around 10% (Liang and others 142; Kang and others 143).

Gene editing has been demonstrated in many cells including cultured human cells and fertilized rodent and non-human zygotes and is easily achieved using a specific single guide RNA (sgRNA) to guide and target the Cas9 protein to the sequence position to be modified (Li and others 119; Shen and others 144; Niu and others 123).

We show that Cas9-mediated gene editing is highly efficient in human 2PN zygotes.

Our work shows the feasibility of the correction of disease-causing genetic defects in human zygotes.

Materials and methods

The Cas9 protein and sgRNAs were diluted into injection buffer (0.25 mM EDTA/10 mM TrisHCl, pH7.4) and incubated for 10 min at 37 °C before injection.

They also gave signed informed consent for use of their immature oocytes and leftover sperm to produce normal (2PN) zygotes for research.

The zygotes were observed under an inverted microscopy 16–18 h after the insemination and the 3PN embryos were selected for experiments.

Genomic DNA from cultured embryos was amplified with the REPLI-g Single Cell Kit (QIAGEN) according to the manufacturer's instructions.

Whole genome sequencing (WGS) and WGS data analysis were performed by Beijing Genomics Institute (BGI).

The amplified DNA was sequenced as paired-end 90-nucleotide reads to a target of 30X haploid coverage on an Illumina HisSeq2000 sequencer.

Results

This RAG1 targeting efficiency is comparable with the reported efficiency of 9/15 (60%) observed for Cas9 mRNA/sgRNA injection into cynomolgus monkey one-cell embryos (Niu and others 123).

Encouraged by the high observed efficiencies of generating an indel mutation (NHEJ editing) at the selected loci using Cas9 protein, we next attempted to determine the HDR efficiency in human 3PN embryos using Cas9 protein.

Having established that CRISPR/Cas9-mediated gene editing (NHEJ and HDR) could be achieved efficiently in human 3PN embryos, we decided to determine the editing efficiencies directly in human 2PN to assess the possibility of using CRISPR/Cas9 in correcting disease-causing mutations.

Further sequencing of embryos #4 and #8 revealed that all the mutant alleles contained an additional C deletion within HBB sgRNA2 targeting site, which cannot be generated by DSB repair by the embryos.

We injected eight 3PN embryos to test the ability of the G6PD sgRNA3 to target the wildtype allele.

Discussion

We demonstrate here that the CRISPR/Cas9 system is quite effective in correcting point mutations in human zygotes.

One could introduce inactivating mutations and even point mutations to probe gene function in early human embryos.

Acknowledgements

A machine generated summary based on the work of Tang, Lichun; Zeng, Yanting; Du, Hongzi; Gong, Mengmeng; Peng, Jin; Zhang, Buxi; Lei, Ming; Zhao, Fang; Wang, Weihua; Li, Xiaowei; Liu, Jianqiao, 2017 in Molecular Genetics and Genomics.

Easi-CRISPR: a robust method for one-step generation of mice carrying conditional and insertion alleles using long ssDNA donors and CRISPR ribonucleoproteins.

https://doi.org/10.1186/s13059-017-1220-4

Abstract-Summary

Transgenic mice generated by random genomic insertion approaches pose problems of unreliable expression, and thus there is a need for targeted-insertion models.

Although CRISPR-based strategies were reported to create conditional and targeted-insertion alleles via one-step delivery of targeting components directly to zygotes, these strategies are quite inefficient.

We describe Easi-CRISPR (Efficient additions with ssDNA inserts-CRISPR), a targeting strategy in which long single-stranded DNA donors are injected with pre-assembled crRNA + tracrRNA + Cas9 ribonucleoprotein (ctRNP) complexes into mouse zygotes.

We show for over a dozen loci that Easi-CRISPR generates correctly targeted conditional and insertion alleles in 8.5–100% of the resulting live offspring.

Easi-CRISPR solves the major problem of animal genome engineering, namely the inefficiency of targeted DNA cassette insertion.

It is versatile, generating both conditional and targeted insertion alleles.

Extended:

These results indicate that Easi-CRISPR can efficiently insert sequences that encode and express reporters, recombinases, and regulatory proteins, and that the technique is applicable to multiple genomic loci.

We anticipate that because of its numerous benefits, including simplicity of design, high efficiency, effectiveness for many genes, and suitability for both low- and high-throughput laboratories, Easi-CRISPR will serve as an effective means of rapidly building mouse Cre-LoxP resources, and for building similar resources for rat and other models in the future.

Background

CRISPR/Cas9-directed genome editing should, in theory, allow for the more rapid generation of floxed alleles in any chosen genetic background, because the editing components can be delivered directly to single-cell mouse zygotes of any strain.

Within months of the first demonstration of CRISPR/Cas9 genome editing to produce small gene disruptions in mammalian cells [108, 119], a proof-of-concept study showed that conditional knockout mice could be generated by homology-directed repair (HDR) following injection of mouse zygotes with five components: two separate single guide RNAs (sgRNAs) targeted to sequences flanking an exon of interest; two single-stranded oligodeoxynucleotide (ssODN) donors, each containing a LoxP site flanked by short (40–80 bases) arms homologous to the desired insertion site; and Cas9 mRNA.

Based on our experience with using ssDNA donors and an sgRNA to insert ~400-base fragments into the mouse genome with high efficiency when assayed at embryonic stages [141], we asked whether longer ssDNA donors and two guide RNAs could be used to generate mice with floxed exons.

Results

Three other pups had partial insertions of the donor cassette: two contained only a single targeted LoxP site and one contained both LoxP insertions, but they were located on separate alleles (in trans).

To determine whether a similar approach could enhance the frequency of HDR with long ssDNA donors, we prepared a crRNA + tracrRNA + Cas9 protein complex using chemically synthesized crRNAs and tracrRNAs designed to cleave Pitx1 in exactly the same sites as the sgRNAs described above.

Of the 20 founders with correctly floxed exons, two contained point mutations in the inserted regions (one each for Pitx1 and Ambra1) that may have derived from enzymatic misincorporation during preparation of the ssDNA donor templates.

Based on the success of Easi-CRISPR for floxing various loci, we asked whether similar efficiencies could be obtained for knock-ins of sequences that encode reporters, recombinases, and transcriptional regulators.

Discussion

Given the multitude of potential undesired products that are possible from the NHEJ repair pathway acting at two Cas9 cleavage sites, the high frequency of recovering correctly floxed alleles at seven different loci by using Easi-CRISPR (8.5–100%) was surprising, as previously described strategies reached a maximum of 16% efficiency [124, 142].

By successfully targeting insertions to six loci we demonstrate that Easi-CRISPR is suitable for generating all such knock-in models.

Easi-CRISPR will also be suitable for generating other types of DNA replacements, such as (1) a set of point mutations spread across a region (e.g., up to 1–2 kb long that can be efficiently inserted), (2) testing regulatory sequences, and (3) replacing short stretches of gene segments or coding sequences from other species (e.g., creating humanized mice).

Conclusions

The Easi-CRISPR strategy we describe here uses simplified CRISPR tools; long ssDNA donors and ctRNPs, and allows the insertion of DNA cassettes into genomes with a very high efficiency.

The method has been used at over a dozen loci revealing robustness, high efficiency and, moreover, versatility as it can create conditional as well as recombinase, reporter, and transcriptional effector knock-in alleles.

Methods

CRISPR guide RNAs were designed using CRISPR.mit.edu, or CHOPCHOP, and were used as annealed two-part synthetic crRNA and tracrRNA molecules for all genes (Alt-R™ CRISPR guide RNAs, Integrated DNA Technologies, Inc. (IDT), Coralville, IA, USA and Genome Craft Type CT, FASMAC, Kanagawa, Japan), and as sgRNAs for Pitx1.

The annealed crRNA and tracrRNA (also known as guide RNA) were diluted in microinjection buffer and mixed with Cas9 protein to obtain ctRNP complexes [143].

One-cell stage fertilized mouse embryos were injected with 5–50 ng/μl Cas9 protein (or 10 ng/μl of Cas9 mRNA; for Pitx1 locus), 5–20 ng/μl of annealed crRNA and tracrRNA (or 10 ng/μl of each sgRNA; for Pitx1 locus) and 5–10 ng/μl of ssDNA.

Each locus-specific experiment was performed by injecting zygotes to generate founders until at least one correctly targeted founder animal was obtained.

Acknowledgements

A machine generated summary based on the work of Quadros, Rolen M.; Miura, Hiromi; Harms, Donald W.; Akatsuka, Hisako; Sato, Takehito; Aida, Tomomi; Redder, Ronald; Richardson, Guy P.; Inagaki, Yutaka; Sakai, Daisuke; Buckley, Shannon M.; Seshacharyulu, Parthasarathy; Batra, Surinder K.; Behlke, Mark A.; Zeiner, Sarah A.; Jacobi, Ashley M.; Izu, Yayoi; Thoreson, Wallace B.; Urness, Lisa D.; Mansour, Suzanne L.; Ohtsuka, Masato; Gurumurthy, Channabasavaiah B, 2017 in Genome Biology.

Genome editing reveals a role for OCT4 in human embryogenesis.

https://doi.org/10.1038/nature24033

Abstract-Summary

We use CRISPR–Cas9-mediated genome editing to investigate the function of the pluripotency transcription factor OCT4 during human embryogenesis.

We identified an efficient OCT4-targeting guide RNA using an inducible human embryonic stem cell-based system and microinjection of mouse zygotes.

Using these refined methods, we efficiently and specifically targeted the gene encoding OCT4 (POU5F1) in diploid human zygotes and found that blastocyst development was compromised.

We conclude that CRISPR–Cas9-mediated genome editing is a powerful method for investigating gene function in the context of human development.

Extended:

This proof of principle lays out a framework for future investigations that could transform our understanding of human biology, thereby leading to improvements in the establishment and therapeutic use of stem cells and in IVF treatments.

Main

The mechanisms that pattern the human embryo are unclear, because of a lack of methods to efficiently perturb gene expression of early lineage specifiers in this species.

To determine whether CRISPR–Cas9 can be used to understand gene function in human preimplantation development, we chose to target POU5F1, a gene encoding the developmental regulator OCT4, as a proof-of-principle.

By using an inducible human ES cell-based CRISPR–Cas9 system and optimizing mouse zygote microinjection techniques, we have identified conditions that allowed us to target POU5F1 efficiently and precisely in human zygotes.

The insights gained from these investigations advance our understanding of human development and suggest that OCT4 has an earlier role in the progression of the human blastocyst compared to the mouse, and therefore that there are distinct mechanisms of lineage specification between these species.

[Section 2]

sgRNA2b was the most efficient at rapidly causing loss of OCT4 protein expression, with only 15.6% of cells retaining detectable OCT4 by day 5 of induction.

[Section 3]

To compare the on-target editing efficiencies and mutation spectrums induced by candidate sgRNAs, we performed a time-course genotypic analysis on cells collected across four days after sgRNA induction.

sgRNA activity in mouse embryos

Only 53% of embryos microinjected with sgRNA2b and Cas9 mRNA exhibited this range of indels.

A greater proportion of blastocysts that formed after sgRNA2b and Cas9 mRNA microinjection had six or more different types of detectable indels (42%) compared to those that formed after microinjection of the sgRNA2b–Cas9 complex (8%).

Having thus determined sgRNA2b to be an efficient and specific guide capable of generating a null mutation of POU5F1 or Pou5f1 in human ES cells and mouse preimplantation embryos, respectively, we next used this guide together with our optimized microinjection technique to target POU5F1 in human preimplantation embryos.

[Section 5]

To test whether OCT4 is required in human embryos, we performed CRISPR–Cas9 editing on thawed in vitro fertilized (IVF) zygotes that were donated as surplus to infertility treatment.

These data suggest that CRISPR–Cas9 targeting does not increase the rate of karyotypic anomalies in human embryos.

These findings suggest that POU5F1 targeting efficiency is high, and that only embryos with partial OCT4 expression are able to progress to the blastocyst stage.

To determine whether there is a high degree of editing in embryos before the onset of OCT4 expression, we microinjected four additional human embryos with the sgRNA2b–Cas9 complex and stopped their development before the eight-cell stage.

With the cleavage-arrested embryos above, these data show that in 45% (five out of eleven) of cleavage stage embryos (either stopped or developmentally arrested), all of the cells analysed from each embryo had no detectable POU5F1 wild-type alleles, indicating high rates of editing.

Loss of OCT4 associated with gene mis-expression

To identify globally which genes might be affected by the loss of OCT4, we microdissected single cells from microinjected embryos at the blastocyst stage.

Coupled with the failure to maintain a fully expanded blastocyst, this finding suggests that the integrity of the trophectoderm may be compromised in OCT4-targeted embryos.

The lack of expression of genes associated with all three lineages in the blastocysts suggests that OCT4-targeted embryos either failed to initiate the expression of these genes or downregulated their expression as development progressed.

This revealed that while cells from OCT4-targeted embryos were progressing towards the transcriptional state of the blastocyst, they were more dispersed and heterogeneous in their gene expression.

Discussion

As the mouse maternal–zygotic Pou5f1-null mutation phenocopies the zygotic-null mutation [144], it is unlikely that persistence of maternal transcripts or proteins compensates for the loss of OCT4 expression, and any additional compensatory mechanisms that may be present in the mouse do not appear to be conserved in the regulation of human development.

The mis-expression of genes associated with all three blastocyst lineages in OCT4-targeted human blastocysts further suggests that OCT4 may have an essential function before this stage.

Inducing POU5F1-null mutations in human embryos slightly later in development, following the onset of EGA, may bypass its earlier critical role and thereby delineate its function in the fully formed blastocyst.

We have developed an optimized approach to target OCT4 in human embryos, thus suggesting that OCT4 has a different function in humans than in mice.

Methods

Genomic DNA from fixed embryos (human and mouse) was isolated using the alkaline lysis method; 25 µl of 50 mM NaOH was added to the sample and incubated at 95 °C for 5 min.

To genotype cells from unfixed Cas9 control or OCT4-targeted human embryos, genomic DNA was isolated from either an individual single cell (1-cell embryos) or following microdissection of multiple individual single-cell samples from each embryo or approximately five cells from trophectoderm biopsies.

The samples were genotyped following whole genome amplification (WGA) using one of the following protocols: (1) For the single cell samples used in the either the modified G&T-seq protocol [145] or isolated solely for genotyping, genomic DNA was amplified using the REPLI-g Single Cell Kit (Qiagen; 150,343) according to the manufacturer's guidelines.

Principal components analysis was performed using the stats (version 3.2.2) R package on a previously published single cell RNA-seq dataset covering different stages of preimplantation development [146] together with our own OCT4-targeted samples and controls.

Acknowledgements.

A machine generated summary based on the work of Fogarty, Norah M. E.; McCarthy, Afshan; Snijders, Kirsten E.; Powell, Benjamin E.; Kubikova, Nada; Blakeley, Paul; Lea, Rebecca; Elder, Kay; Wamaitha, Sissy E.; Kim, Daesik; Maciulyte, Valdone;

Kleinjung, Jens; Kim, Jin-Soo; Wells, Dagan; Vallier, Ludovic; Bertero, Alessandro; Turner, James M. A.; Niakan, Kathy K, 2017 in Nature.

RNA targeting with CRISPR–Cas13.

https://doi.org/10.1038/nature24049

Main

We next evaluated the ability of LwaCas13a to cleave transcripts in mammalian cells.

We transfected the LwaCas13a expression vector, guide vector, and dual-luciferase construct into HEK293FT cells and measured luciferase activity 48 h after transfection.

To compare LwaCas13a knockdown with RNAi, we selected the top three performing guides against Gluc and Cluc and compared them to position-matched shRNAs.

To comprehensively search for off-target effects of LwaCas13a knockdown, we performed transcriptome-wide mRNA sequencing.

These results show that LwaCas13a can be reprogrammed with guide RNAs to effectively knockdown or bind transcripts in mammalian cells.

LwaCas13a knockdown is comparable to RNAi knockdown efficiency, but with substantially reduced off-targets, making it potentially well-suited for therapeutic applications.

We anticipate that there will be additional applications for LwaCas13a and dLwaCas13a, such as genome-wide pooled knockdown screening, interrogation of lncRNA and nascent transcript function, pulldown assays to study RNA–protein interactions, translational modulation, and RNA base editing.

Methods

Post-transformation, cells were recovered at 37 °C in 500 µl of SOC (ThermoFisher Scientific) per biological replicate for 1 h, plated on bio-assay plates (Corning) with LB-agar (Affymetrix) supplemented with 100 µg µl^{-1} ampicillin and 25 µg µl^{-1} chloramphenicol, and incubated at 37 °C for 16 h. Colonies were then harvested by scraping, and plasmid DNA was purified with NuceloBond Xtra EF (Macherey–Nagel) for subsequent sequencing.

For gene expression experiments in mammalian cells, cell harvesting and reverse transcription for cDNA generation was performed using a previously described modification [147] of the commercial Cells-to-Ct kit (Thermo Fisher Scientific) 48 h after transfection.

For RNA immunoprecipitation experiments, HEK293FT cells were plated in six-well plates and transfected with 1.3 µg of dLwaCas13a expression plasmid and 1.7 µg of guide plasmid, with an additional 150 ng of reporter plasmid for conditions involving reporter targeting.

Acknowledgements

A machine generated summary based on the work of Abudayyeh, Omar O.; Gooten-berg, Jonathan S.; Essletzbichler, Patrick; Han, Shuo; Joung, Julia; Belanto, Joseph J.; Verdine, Vanessa; Cox, David B. T.; Kellner, Max J.; Regev, Aviv; Lander, Eric S.; Voytas, Daniel F.; Ting, Alice Y.; Zhang, Feng, 2017 in Nature.

Simultaneous single-cell profiling of lineages and cell types in the vertebrate brain.

https://doi.org/10.1038/nbt.4103

Main

The reconstruction of developmental trajectories from scRNA-seq data requires deep sampling of intermediate cell types and states [148–152] and is unable to capture the lineage relationships of cells.

Lineage tracing methods using viral DNA barcodes, multicolor fluorescent reporters or somatic mutations have not been coupled to single-cell transcriptome readouts, hampering the simultaneous large-scale characterization of cell types and lineage relationships [153, 154].

We develop an approach that extracts lineage and cell type information from a single cell.

We applied scGESTALT to the zebrafish brain and identified more than 100 different cell types and created lineage trees that help reveal spatial restrictions, lineage relationships, and differentiation trajectories during brain development.

scGESTALT can be applied to most multicellular systems to simultaneously uncover cell type and lineage for thousands of cells.

Results

To classify each cluster, we systematically compared differentially expressed genes with prior annotations of gene expression in specific cell types or brain regions in the literature and the ZFIN database [155, 156].

In our initial implementation of GESTALT, all editing reagents (Cas9 protein and sgRNAs) were injected into one-cell-stage embryos, thus centering barcode editing on pre-gastrulation stages [157].

To test this technology (scGESTALT), we performed early and late editing at the one-cell stage and at 30 hpf and dissected whole brains at 23–25 dpf.

ScGESTALT barcodes overlapped nearly all broadly defined cell types (62/63 broad clusters), indicating that the lineage transgene is widely expressed in the brain.

These results establish scGESTALT as a technology that enables the simultaneous recovery of edited barcodes and transcriptomes from single cells.

The zebrafish brain maintains widespread neurogenic activity [158], raising the possibility that scGESTALT could generate edited barcodes that are still shared between progenitors and differentiated cells at the time of cell isolation.

Discussion

The recent application of DNA editing technologies to introduce cumulative, combinatorial, permanent, and heritable changes into the genome has enabled the reconstruction of lineage trees at unprecedented scales but has been limited by the lack of high-resolution cell type information and the restriction of editing to early embryogenesis [157, 159, 160].

We begin to overcome these limitations by establishing a system for expressing both Cas9 and sgRNAs after zygotic activation, thus enabling early and late editing and applying scRNA-seq to identify both the identity and lineage of cells.

We apply our technology, scGESTALT, to zebrafish brain development and establish its potential to simultaneously define cell types and their lineage relationships at a large scale.

It is now feasible to define dozens of cell types by profiling tens of thousands of cells from tissues such as spinal cord, liver, or skin using scRNA-seq and then use barcode editing to mark thousands of cells and reveal their lineage relationships.

Methods

Wild-type and early- and late-edited 23–25 dpf zebrafish brains were similarly processed for inDrops single-cell transcriptome barcoding [161, 162] except that two-time point-edited zebrafish were first heat-shocked for 45 min at 37 °C to induce scGESTALT barcode mRNA expression.

V3 inDrops libraries are sequenced with standard Illumina sequencing primers in which the biological read is from paired end read1, cell barcodes are from paired end read2 and index read1, and library sample index is from index read2.

Sequencing adapters, sample indexes, and flow cell adapters were incorporated as described for the V3 transcriptome libraries.

For both scGESTALT and transcriptome libraries, error-corrected cell barcode sequences were retained for each cell to enable direct comparisons of transcript and lineage information in downstream steps.

Sequencing data from genomic DNA and inDrops scGESTALT libraries were processed with a custom pipeline (https://github.com/aaronmck/SC_GESTALT) as previously described [157] with the following modifications.

Acknowledgements

A machine generated summary based on the work of Raj, Bushra; Wagner, Daniel E; McKenna, Aaron; Pandey, Shristi; Klein, Allon M; Shendure, Jay; Gagnon, James A; Schier, Alexander F, 2018 in Nature Biotechnology.

Simultaneous lineage tracing and cell-type identification using CRISPR–Cas9-induced genetic scars.

https://doi.org/10.1038/nbt.4124

Main

We set out to analyze the data at higher resolution and reconstruct lineage trees on the level of single cells instead of cell types.

Scar dropouts meant that we did not have full lineage information about every single cell.

We cannot, therefore, expect to find exact correspondence of early lineage trees for all cell types in different animals.

Related single-cell lineage tracing methods based on CRISPR–Cas9 technology have recently been used to study brain development as well as the clonal history of different organ systems in the zebrafish [163, 164].

An important advantage of CRISPR–Cas9 lineage tracing compared to competing technologies, such as viral barcoding and other inducible sequence-based lineage tracing methods, is the ability to move beyond clonal analysis and to computationally reconstruct full lineage trees on the single-cell level.

Methods

We compared the scar sequences found within a cell to each other.

We filtered out sequences that had a Hamming distance of 2 or less to another scar sequence in the same cell that occurred in at least eight times as many reads.

Scar sequences in the same cell that were one Hamming distance apart but had a read ratio less than 8 were tested on three criteria if both of them occurred at least twice in the scar library: 1 Do both scars have more than one transcript?

For each cell type we determined the distribution of different scars seen per cell and set a maximum number of scars a cell of that type can have.

To account for the slightly different sequencing read structure of single-cell and bulk scar detection (see above), we considered only the nucleotides that are shared between the two approaches, and we assigned the bulk scar probabilities to single-cell scars accordingly.

Acknowledgements

A machine generated summary based on the work of Spanjaard, Bastiaan; Hu, Bo; Mitic, Nina; Olivares-Chauvet, Pedro; Janjuha, Sharan; Ninov, Nikolay; Junker, Jan Philipp, 2018 in Nature Biotechnology.

A CRISPR–Cas9 gene drive targeting doublesex causes complete population suppression in caged Anopheles gambiae mosquitoes.

https://doi.org/10.1038/nbt.4245

Main

CRISPR–Cas9 nucleases have been applied in gene drive constructs to target endogenous sequences of the human malaria vectorsA. gambiae and A. stephensi with the objective of vector control [165, 166].

According to mathematical modeling, suppression of A. gambiae mosquito reproductive capability can be achieved using gene drive systems targeting haplosufficient
female fertility genes [167, 168] or by introducing a sex distorter on the Y chromosome in the form of a nuclease designed to shred the X chromosome during meiosis,
an approach known as Y-drive [168–170].

A gene drive designed to disrupt the A. gambiae fertility gene AGAP007280
initially increased in frequency, but the selection of nuclease-resistant, functional
variants that could be detected as early as generation 2 completely blocked the spread
of the drive [166].

Gene drive targets with functional or structural constraints that might prevent the
development of resistant variants could offer a route to successful population control.

Results

The drastic phenotype of dsxF$^{-/-}$ in females indicates that exon 5 of dsx has a
fundamental role in the previously poorly understood sex differentiation pathway of
A. gambiae mosquitoes and suggested that its sequence might represent a suitable
target for gene drives designed for population suppression.

To test this hypothesis, we mixed caged wild-type mosquito populations with
heterozygous individuals carrying the dsxF^{CRISPRh} allele and monitored progeny
at each generation to assess the spread of the drive and to quantify effect(s) on
reproductive output.

We started the experiment in two replicate cages, each with an initial drive allele
frequency of 12.5% (300 wild-type female mosquitoes with 150 wild-type male
mosquitoes and 150 dsxF^{CRISPRh}/ + male individuals).

The observation that heterozygous dsxF^{CRISPRh}/ + females are fertile and produce
almost 100% inheritance of the drive might indicate that most of the germ cells in
these females are homozygous and, unlike somatic cells, do not undergo autonomous
dsx-mediated sex commitment [171].

Discussion

The development of a gene drive capable of collapsing a human malaria vector population to levels that cannot support malaria transmission is a long-sought scientific
and technical goal [172].

Invasion of the drive in transformer was rapidly compromised by the accumulation
of large numbers of functional and nonfunctional resistant alleles [173].

Our doublesex gene drive now needs to be rigorously evaluated in large confined
spaces that more closely mimic native ecological conditions, in accordance with the
recommendations of the US National Academy of Sciences [174].

Competition for resources or mating success may disproportionately affect individuals harboring the gene drive, resulting in invasion dynamics substantially
different from those observed in insectary cage experiments.

Methods

Generation of the HDR-mediated dsxF^{-} allele was confirmed using primers
binding the integrated cassette (GFP-F and 3xP3-R) and the neighboring genomic

integration site, external to the sequence included on the homology arms (dsxin3-F and dsxex6-R).

Gametes (W, D or R) from W/D females and W/D, D/R and D/D males carry nuclease that is transmitted to the zygote and acts in the embryo in somatic cells to reduce fitness if wild-type alleles are present, so that W/W, W/R and W/D females have fitness w10, w01, w11 or 1, depending on whether nuclease was derived from a transgenic mother, father, both or neither.

To W and R gametes that are derived from parents that have no drive allele and therefore have no deposited nuclease, gametes from W/D females and W/D, D/R and D/D males carry nuclease that is transmitted to the zygote, and these are denoted W^*, D^* and R^*.

Acknowledgements

A machine generated summary based on the work of Kyrou, Kyros; Hammond, Andrew M; Galizi, Roberto; Kranjc, Nace; Burt, Austin; Beaghton, Andrea K; Nolan, Tony; Crisanti, Andrea, 2018 in Nature Biotechnology.

Anti-CRISPR-mediated control of gene editing and synthetic circuits in eukaryotic cells.

https://doi.org/10.1038/s41467-018-08158-x

Abstract-Summary

We characterize a panel of anti-CRISPR molecules for expanded applications to counteract CRISPR-mediated gene activation and repression of reporter and endogenous genes in various cell types.

We demonstrate that cells pre-engineered with anti-CRISPR molecules become resistant to gene editing, thus providing a means to generate "write-protected" cells that prevent future gene editing.

Our work suggests that anti-CRISPR proteins should serve as widely applicable tools for synthetic systems regulating the behavior of eukaryotic cells.

Introduction

CRISPR systems, a form of prokaryotic adaptive immunity, have been widely repurposed for biotechnological applications, including genome editing and gene expression regulation in prokaryotic and eukaryotic organisms [175, 176].

Interest in synthetic circuits implemented by CRISPR systems [177, 178] has grown due to the adaptability of CRISPR-based gene regulation.

The complexity of implementable circuits is limited by the types of control nodes that can be wired to control CRISPR systems.

The broad extent as to whether Acrs can be used as tools to provide temporal, inhibitory control of CRISPR genome editors and nuclease-deactivated Cas9 (dCas9) genome regulators (both activation and repression) in different eukaryotic cells remains to be characterized.

We present a more complete characterization of Acr activity in a range of contexts and establish the basis for biotechnological applications involving the use of Acrs for controlling CRISPR activity in mammalian cells.

Results

CRISPR-based regulation of gene expression involves the use of a dCas9 with target sequence specified by a single guide RNA (sgRNA).

To do this, we systematically assessed the efficacy of a panel of 5 Acrs (AcrIIC1, AcrIIA1, AcrIIA2, AcrIIA3, AcrIIA4) targeting Class II CRISPR systems [179, 180] to alter gene expression changes induced by CRISPRa and CRISPRi.

We further explored the utility of the best-performing AcrIIA4 in controlling gene regulation.

This suggests AcrIIA4 works for controlling CRISPR-based gene regulation in diverse mammalian cell types as a general tool.

These assays in concert suggest the following results: AcrIIA1 is a strong inhibitor of Cas9 editing in yeast but not dCas9-based gene regulation, implying a possible mechanism of inhibition at the editing level (perhaps akin to a previously reported mechanism for a II-C Acr [181]); and AcrIIA2 demonstrates stronger apparent activity for gene editing and gene regulation in yeast than in mammalian cells (consistent with other reports [182, 183]).

We further engineered Acrs for inducible control of gene expression.

Discussion

Our results demonstrated that AcrIIA4 is a potent regulator of (d)Cas9 activity in a wide variety of contexts (reporter or endogenous genes) and cell types (HEK293T, hiPSC, and yeast).

This, combined with other reports [179, 180, 184], contributes to a picture on the use of Acrs as an additional layer of control over (d)Cas9 activity in eukaryotic cells.

Further, the combination of CRISPRa and CRISPRi assays (in addition to editing assays) allows for distinguishing between specific anti-CRISPR activity and potential cytotoxic effects and revealed that other Acrs demonstrate varying activity depending on the organismal and CRISPR activity (knockout versus gene regulation) context.

The use of Acrs in regulating CRISPR activity should allow for the generation of more advanced dynamic control over gene regulation.

The level at which Acrs operate is distinct from other methods of inducible control [185, 186], allowing for multilayered logical control over dCas9-based gene regulation.

Methods

Plasmids and cell lines were generated using standard molecular cloning techniques.

Lentivirus for generating reporter cell lines and sgRNA transduction was packaged using wild-type HEK293T (Clontech).

The cell line containing the IFFL was sorted for negative mCherry expression, then transfected with a plasmid bearing tetracycline-controlled transactivator (tTA) and sorted for positive expression 2 days post transfection.

For DD-Shield1 experiments, medium of cells was supplemented with appropriate amount of Shield1 ligand immediately following transfection.

For gain-of-function editing experiments, a HEK293T reporter cell line with out-of-frame split-GFP construct was transfected with Cas9 and sgRNA plasmid with and without plasmid containing AcrIIA4.

Three days after transfection, cells were analyzed via flow cytometry without gating for presence of plasmid.

For write-protection editing experiments, HEK293T cells were transiently transfected with plasmid bearing Cas9-eGFP and sgRNA, as above, or Cas9 protein (IDT or Synthego) with synthetic modified sgRNA (Synthego) via electroporation (Invitrogen Neon or Amaxa Nucleofection).

Acknowledgements

A machine generated summary based on the work of Nakamura, Muneaki; Srinivasan, Prashanth; Chavez, Michael; Carter, Matthew A.; Dominguez, Antonia A.; La Russa, Marie; Lau, Matthew B.; Abbott, Timothy R.; Xu, Xiaoshu; Zhao, Dehua; Gao, Yuchen; Kipniss, Nathan H.; Smolke, Christina D.; Bondy-Denomy, Joseph; Qi, Lei S, 2019 in Nature Communications.

Super-Mendelian inheritance mediated by CRISPR–Cas9 in the female mouse germline.

https://doi.org/10.1038/s41586-019-0875-2

Abstract-Summary

Highly efficient gene drive systems have recently been developed in insects, which leverage the sequence-targeted DNA cleavage activity of CRISPR–Cas9 and endogenous homology-directed repair mechanisms to convert heterozygous genotypes to homozygosity [165–187, 187, 188].

We use an active genetic element that encodes a guide RNA, which is embedded in the mouse tyrosinase (Tyr) gene, to evaluate whether targeted gene conversion can occur when CRISPR–Cas9 is active in the early embryo or in the developing germline.

Although Cas9 efficiently induces double-stranded DNA breaks in the early embryo and male germline, these breaks are not corrected by homology-directed repair.

Cas9 expression limited to the female germline induces double-stranded breaks that are corrected by homology-directed repair, which copies the active genetic element from the donor to the receiver chromosome and increases its rate of inheritance in the next generation.

Main

We first tested whether Cas9 in the female germline could promote copying of the $Tyr^{CopyCat}$ element onto the receiver chromosome by crossing F_3 female mice of each Vasa-cre lineage to CD-1 (Tyr^{null}) males.

To early embryonic expression of cas9, we observed that the Tyr^{CopyCat} transgene was copied to the Tyr^{ch}-marked receiver chromosome in both Vasa-cre;Rosa26-LSL-cas9 and Vasa-cre;H11-LSL-cas9 lineages.

The observed efficiency differed between genotypes; three out of five females of the Vasa-cre;Rosa26-LSL-cas9 lineage and all five females of the Vasa-cre;H11-LSL-cas9 lineage transmitted a Tyr^{ch}-marked chromosome that contained a Tyr^{CopyCat} insertion to at least one offspring.

The average observed copying rate of 44% using the most efficient genetic strategy in females (Vasa-cre;H11-LSL-cas9) combined ultra-tightly linked tyrosinase mutations such that 22.5% of all offspring inherited a chromosome with both alleles, which would not be possible through Mendelian inheritance.

Methods

To detect gene conversion events, we used genetic linkage rather than a statistical test of inheritance greater than 50% expected by Mendelian segregation.

The receiver chromosome was marked with a SNP (Tyr^{ch}) located approximately 9.1 kb from the target site for gene conversion.

Acknowledgements

A machine generated summary based on the work of Grunwald, Hannah A.; Gantz, Valentino M.; Poplawski, Gunnar; Xu, Xiang-Ru S.; Bier, Ethan; Cooper, Kimberly L, 2019 in Nature.

CasX enzymes comprise a distinct family of RNA-guided genome editors.

https://doi.org/10.1038/s41586-019-0908-x

Abstract-Summary

The RNA-guided CRISPR-associated (Cas) proteins Cas9 and Cas12a provide adaptive immunity against invading nucleic acids, and function as powerful tools for genome editing in a wide range of organisms.

We reveal the underlying mechanisms of a third, fundamentally distinct RNA-guided genome-editing platform named CRISPR–CasX, which uses unique structures for programmable double-stranded DNA binding and cleavage.

Eight cryo-electron microscopy structures of CasX in different states of assembly with its guide RNA and double-stranded DNA substrates reveal an extensive RNA scaffold and a domain required for DNA unwinding.

Main

Sequence analysis of CasX revealed no similarity to other CRISPR–Cas enzymes, except for the presence of a RuvC nuclease domain that is also found in both Cas9 and Cas12a enzyme families, transposases, and recombinases [189].

Phylogenetic, biochemical and structural data show that CasX contains domains distinct from—but analogous to—those found in Cas9 and Cas12a, as well as novel

RNA and protein folds; thus establishing the CasX enzyme family as the third CRISPR–Cas platform that is effective for genetic manipulation.

Distinct conformational states observed for CasX suggest an ordered non-target- and target-strand cleavage mechanism that may explain how CRISPR–Cas enzymes with a single active site, such as Cas12a, achieve double-stranded DNA (dsDNA) cleavage [190–192].

The small size of CasX (<1,000 amino acids), its DNA cleavage characteristics, and its derivation from non-pathogenic microorganisms offer important advantages over other CRISPR–Cas genome-editing enzymes.

Reconstituting RNA-guided CasX cutting of dsDNA

This mode of dsDNA cleavage is consistent with the staggered cuts to DNA observed for Cas12a and Cas12b (also known as C2c1), which represent other CRISPR–Cas enzymes that use a single RuvC active site for DNA cleavage [190, 192, 193].

To test whether CasX displays similar non-specific activity, single-stranded phage DNA was incubated with DpbCasX–guide RNA complexes that target a separate, unrelated dsDNA substrate.

CasX enacts genome silencing and editing

To determine whether the RNA-guided DNA cutting activity of CasX can be harnessed for programmed genome targeting, DpbCasX and its sgRNA were expressed in E. coli using a guide sequence that is complementary to an integrated reporter in the genome of bacterial strain MG1655 [194, 195].

We next tested whether CasX is capable of inducing cleavage and gene editing of mammalian genomes.

We explored the effect of the concentration of the plasmid that encodes CasX sgRNA on the extent of genome editing.

These results demonstrate that CasX belongs to a third, distinct class of CRISPR systems that is capable of targeted genomic regulation and editing, and motivated experiments aimed at determining the structural and mechanistic basis for these activities.

CasX has a unique domain composition

Although structural alignment of the entire modelled polypeptide chain revealed some similarity between CasX and Cas12a (Lachnospiraceae bacterium Cpf1 (LbCpf1), RCSB Protein Data Bank (PDB) code 5XUU); z-score 15.1, with a root mean square deviation (r.m.s.d.) value of 5.3 Å for 671 aligned residues) [196], a more detailed analysis of the domains showed that—as expected—this similarity results from the RuvC domain and oligonucleotide-binding domain (OBD); alignment of RuvC with LbCpf1 PDB 5XUT gives a z-score of 13.8 and an r.m.s.d.

We propose that Nuc and similar domains should be renamed TSL to better explain the activity that we postulate to be target-strand placement in the RuvC active site [192].

Cas12a Nuc domain, amino acids Arg1226 and Asp1235 aid target-strand cleavage, and an Arg1226Ala mutation produced an Acidaminococcus sp.

Two conformations enable CasX dsDNA cleavage

This preference suggests that non-target-strand DNA is cleaved by the RuvC domain first, followed by displacement and target-strand cleavage.

The CasX NTSB domain is required for DNA unwinding

These results suggest that the NTSB domain is responsible for initiating and/or stabilizing DNA duplex unwinding by CasX. This finding also hints at the interesting possibility that the self-contained NTSB domain could be introduced into or acquired by other enzymes to assist with or stabilize dsDNA binding.

Conclusions

The compact size, dominant RNA content and minimal trans-cleavage activity of CasX differentiate this enzyme family from Cas9 and Cas12a, and provide opportunities for therapeutic delivery and safety that may offer important advantages relative to existing genome-editing technologies.

Methods

Cells were collected, re-suspended in Ni buffer A (500 mM sodium chloride, 50 mM HEPES, pH 7.5, 10% glycerol, 0.5 mM TCEP) and frozen at $-$ 80 °C For wild-type CasX protein preparation, cells were thawed, diluted twice with Ni buffer A, followed by addition of PMSF (final concentration 0.5 mM), and 3 tablets of Roche protease inhibitor cocktail per 100 ml of cell suspension.

For CasX(D672A/E769A/D935A), the purification was similar, except that dialysis buffer was: 300 mM sodium chloride, 50 mM HEPES, pH 7.5, 10% glycerol, 0.5 mM TCEP and size-exclusion buffer was 300 mM potassium chloride, 50 mM HEPES, pH 7.5, 10% glycerol, 0.5 mM TCEP, and all the protein eluted as a single well-folded protein peak on heparin column.

Apo-CasX in a buffer containing 20 mM HEPES, pH 7.5, 500 mM NaCl, 1 mM DTT and 5% glycerol was used for cryo-EM sample preparation following the same sample protocol used for CasX complexes.

Acknowledgements

A machine generated summary based on the work of Liu, Jun-Jie; Orlova, Natalia; Oakes, Benjamin L.; Ma, Enbo; Spinner, Hannah B.; Baney, Katherine L. M.; Chuck, Jonathan; Tan, Dan; Knott, Gavin J.; Harrington, Lucas B.; Al-Shayeb, Basem; Wagner, Alexander; Brötzmann, Julian; Staahl, Brett T.; Taylor, Kian L.; Desmarais, John; Nogales, Eva; Doudna, Jennifer A, 2019 in Nature.

Multiplexed detection of proteins, transcriptomes, clonotypes and CRISPR perturbations in single cells.

https://doi.org/10.1038/s41592-019-0392-0

Abstract-Summary

We describe expanded CRISPR-compatible cellular indexing of transcriptomes and epitopes by sequencing (ECCITE-seq) for the high-throughput characterization of at least five modalities of information from each single cell.

Main

Several approaches have recently been reported that allow detection of CRISPR-mediated perturbations along with the transcriptome of single cells using specialized vectors that link the expression of single-guide RNAs (sgRNAs) to separate transcripts that can be captured by standard scRNA-seq methods [197–200].

We and others have layered detection of proteins on top of scRNA-seq to enable integration of robust and well-characterized protein markers with unbiased transcriptomes of single cells [201, 202].

Our method, cellular indexing of transcriptomes and epitopes by sequencing (CITE-seq) is compatible with oligo-dT based scRNA-seq approaches and enables simultaneous protein detection using DNA oligo-labeled antibodies against cell surface markers.

We extend the use of CITE-seq and the related Cell Hashing method for multiplexing and doublet detection [203], to 5′ capture-based scRNA-seq methods, exemplified by the 10 × Genomics 5P/V(D)J system, allowing the detection of surface proteins together with the scRNA-seq and clonotype features.

Methods

Stained and washed cells were loaded into a 10 × Genomics single-cell V(D)J workflow and processed according to the manufacturer's instructions with the following modifications: (1) 12 pmol of an reverse-transcription primer complementary to sgRNA scaffold sequences was spiked into the reverse-transcription reaction (only when sgRNA capture was desired).

The complementary DNA fraction was processed according to the 10 × Genomics Single-Cell V(D)J protocol to generate the transcriptome library and the TCR α/β library.

cDNA and TCR (α/β and γ/δ) enriched libraries were further processed according to the 10 × Genomics Single-Cell V(D)J protocol. (4) An additional 1.4 × reaction volume of SPRI beads was added to the protein-tag/hashtag/guide-tag fraction from step 3, to bring the ratio up to 2.0 × . Beads were washed with 80% ethanol, eluted in water and an additional round of 2.0 × SPRI performed to remove excess single-stranded oligonucleotides carried over from the cDNA amplification reaction.

Acknowledgements

A machine generated summary based on the work of Mimitou, Eleni P.; Cheng, Anthony; Montalbano, Antonino; Hao, Stephanie; Stoeckius, Marlon; Legut, Mateusz; Roush, Timothy; Herrera, Alberto; Papalexi, Efthymia; Ouyang,

Zhengqing; Satija, Rahul; Sanjana, Neville E.; Koralov, Sergei B.; Smibert, Peter, 2019 in Nature Methods.

Transposon-encoded CRISPR–Cas systems direct RNA-guided DNA integration.

https://doi.org/10.1038/s41586-019-1323-z

Abstract-Summary

Conventional CRISPR–Cas systems maintain genomic integrity by leveraging guide RNAs for the nuclease-dependent degradation of mobile genetic elements, including plasmids and viruses.

We describe a notable inversion of this paradigm, in which bacterial Tn7-like transposons have co-opted nuclease-deficient CRISPR–Cas systems to catalyse RNA-guided integration of mobile genetic elements into the genome.

Programmable transposition of Vibrio cholerae Tn6677 in Escherichia coli requires CRISPR- and transposon-associated molecular machineries, including a co-complex between the DNA-targeting complex Cascade and the transposition protein TniQ. Integration of donor DNA occurs in one of two possible orientations at a fixed distance downstream of target DNA sequences, and can accommodate variable length genetic payloads.

Main

The evolution of CRISPR–Cas is intimately linked to the large reservoir of genes provided by mobile genetic elements, with core enzymatic machineries involved in both new spacer acquisition (Cas1) and RNA-guided DNA targeting (Cas9 and Cas12) derived from transposable elements [49, 204–208].

We were inspired by a recent report that described a class of bacterial Tn7-like transposons encoding evolutionarily linked CRISPR–Cas systems and proposed a functional relationship between RNA-guided DNA targeting and transposition [209].

We therefore hypothesized that transposon-encoded CRISPR–Cas systems have been repurposed for a role other than adaptive immunity, in which RNA-guided DNA targeting is leveraged for a novel mode of transposon mobilization.

Beyond revealing an elegant mechanism by which mobile genetic elements have hijacked RNA-guided DNA targeting for their evolutionary success, our work highlights an opportunity for facile, site-specific DNA insertion without requiring homologous recombination.

Cascade directs site-specific DNA integration

The Tn7 transposon contains characteristic left- and right-end sequences and encodes five tns genes, tnsA–tnsE, which collectively encode a heteromeric transposase: TnsA and TnsB are catalytic enzymes that excise the transposon donor via coordinated double-strand breaks; TnsB, a member of the retroviral integrase superfamily, catalyses DNA integration; TnsD and TnsE constitute mutually exclusive targeting factors that specify DNA insertion sites; and TnsC is an ATPase that communicates between TnsAB and TnsD or TnsE [210].

Given the presence of discrete bands, it appeared that integration was occurring at a set distance from the target site, and Sanger and next-generation sequencing (NGS) analyses revealed that more than 95% of integration events for crRNA-1 occurred 49 base pairs (bp) from the 3' edge of the target site.

These experiments demonstrate transposon integration downstream of genomic target sites complementary to guide RNAs.

Protein requirements of RNA-guided DNA integration

In E. coli, site-specific transposition requires attTn7 binding by EcoTnsD, followed by interactions with the EcoTnsC regulator protein to directly recruit the EcoTnsA-TnsB-donor DNA [211].

Given the essential nature of tniQ (a tnsD homologue) in RNA-guided transposition, and its location within the cas8-cas7-cas6 operon, we envisioned that the Cascade complex might directly bind TniQ and thereby deliver it to genomic target sites.

To determine whether specific TniQ–Cascade interactions are required, or whether TniQ could direct transposition adjacent to generic R-loop structures or via artificial recruitment to DNA, we used Streptococcus pyogenes Cas9 (SpyCas9) [83] and Pseudomonas aeruginosa Cascade (PaeCascade) [212] as orthogonal RNA-guided DNA-targeting systems.

Donor requirements of RNA-guided DNA integration

Shorter transposons containing right-end truncations were integrated more efficiently, accompanied by a notable change in the orientation bias.

Future efforts will be required to explore how transposition is affected by vector design, to what extent transposon end mutations are tolerated, and whether rational engineering allows for integration of larger cargos and/or greater control over integration orientation.

Guide RNA and target DNA requirements

In the absence of any structural data, we realized that we could investigate whether TniQ may be positioned near the PAM-distal end of the R-loop by testing engineered crRNAs that contain spacers of variable lengths.

Previous work with E. coli Cascade has demonstrated that crRNAs with extended spacers form complexes that contain additional Cas7 subunits [213], which would increase the distance between the PAM-bound Cas8 and the Cas6 at the other end of the R-loop.

Although more experiments are required to deduce the underlying mechanisms that explain this bimodal distribution, as well as the insertion site distribution observed for other extended crRNAs, these data, together with the mismatch panel, provide further evidence that TniQ is tethered to the PAM-distal end of the R-loop structure.

Programmability and genome-wide specificity

Our experiments thus far specifically interrogated genomic loci containing the anticipated integration products, and it therefore remained possible that non-specific integration was simultaneously occurring elsewhere, either at off-target genomic sites bound by Cascade, or independently of Cascade targeting.

These experiments highlight the high degree of intrinsic programmability and genome-wide integration specificity directed by transposon-encoded CRISPR–Cas systems.

Discussion

Transposases and integrases are generally thought to mobilize their specific genetic payloads either by integrating randomly, with a low degree of sequence specificity, or by targeting specialized genomic loci through inflexible, sequence-specific homing mechanisms [214].

We have discovered a fully programmable integrase, in which the DNA insertion activity of a heteromeric transposase from V. cholerae is directed by an RNA-guided complex known as Cascade, the DNA-targeting specificity of which can be easily tuned.

Although future experiments will be necessary to determine whether these systems also possess RNA-guided DNA integration activity, the bioinformatic evidence points to a more pervasive functional coupling between CRISPR–Cas systems and transposable elements than previously appreciated.

The ability to INsert Transposable Elements by Guide RNA-Assisted TargEting (INTEGRATE) offers an opportunity for site-specific DNA integration that would obviate the need for double-strand breaks in the target DNA, homology arms in the donor DNA, and host DNA repair factors.

Methods

Reactions were prepared in 384-well clear/white PCR plates (BioRad), and measurements were performed on a CFX384 Real-Time PCR Detection System (BioRad) using the following thermal cycling parameters: polymerase activation and DNA denaturation (98 °C for 2.5 min), 40 cycles of amplification (98 °C for 10 s, 62 °C for 20 s), and terminal melt-curve analysis (65–95 °C in 0.5 °C per 5 s increments).

E. coli BL21 (DE3) cells containing one or both plasmids were grown in 2xYT medium with the appropriate antibiotic(s) at 37 °C to $OD_{600} = 0.5$–0.7, at which point IPTG was added to a final concentration of 0.5 mM and growth was allowed to continue at 16 °C for an additional 12–16 h. Cells were harvested by centrifugation at 4,000 g for 20 min at 4 °C.

Acknowledgements

A machine generated summary based on the work of Klompe, Sanne E.; Vo, Phuc L. H.; Halpin-Healy, Tyler S.; Sternberg, Samuel H, 2019 in Nature.

Rapid and sensitive exosome detection with CRISPR/Cas12a.

https://doi.org/10.1007/s00216-019-02211-4

Abstract-Summary

We report here a method for exosome detection based on the CD63 aptamer and clustered regular interspaced short palindromic repeats (CRISPR)/Cas12a system.

This method consists mainly of exosomal membrane protein recognition based on the CD63 aptamer and signal amplification based on CRISPR/Cas12a.

The CD63 aptamer, as an easily adaptable nucleic acid strand, is responsible for the conversion of the amounts of exosomes into nucleic acid detection, whereas CRISPR/Cas12a is responsible for highly specific nucleic acid signal amplification.

This method provides a highly sensitive and specific method for the detection of exosomes and offers an avenue toward future exosome-based diagnosis of diseases.

Introduction

Proteins play critical roles in maintaining cell recognition and structural reinforcement of the exosomal membrane.

Recent reports revealed that a group of proteins are overexpressed on various cell-derived exosomes, such as cytoplasmic proteins, transmembrane proteins, and Rab [215].

Although traditional immunoaffinity-based techniques such as western blotting, enzyme-linked immunosorbent assay, and flow cytometry have been used for exosome analysis, they are limited by some shortcomings [216, 217].

Various biosensors were also reported to achieve ultrasensitive detection of exosomes through membrane proteins (CD63 and CD9), including by colorimetry, immunofluorescence, surface-enhanced Raman scattering, electrochemistry, and microfluidics [218–220].

We report a rapid and sensitive exosome detection technique based on CD63 aptamer and trans cleavage by CRISPR/Cas12a.

In our design, the CD63 aptamer, partially blocked by complementary DNA strands (blocker), is used to capture CD63-bearing exosomes.

The combination of exosomes and CD63 aptamer triggers a conformational change of CD63 aptamer to release the blocker.

Experimental

Dynabeads MyOne streptavidin magnetic beads (SMBs), Dulbecco's modified Eagle medium, and fetal bovine serum were purchased from Thermo Fisher Scientific (USA).

Exosomes secreted by A549 cells were obtained by ultracentrifugation of cell culture supernatant.

The exosome-containing supernatant was then filtered by a 0.22-μm filter, followed by ultracentrifugation at 100,000 g for 150 min at 4 °C to precipitate exosomes.

The surfaces of SMBs were labeled with Apt63 as follows, First, 5 μl CD63 aptamer and 7.5 μl blocker were mixed with 37.5 μl 1 × PBS and prehybridized with a PCR machine (90 °C for 5 min, cooled to 4 °C at a speed of 4 °C/min) to form Apt63.

Apt63 (10 μl) was added to the diluted SMB solution and mixed for 30 min at room temperature, followed by magnetic separation of the Apt63-labeled SMBs.

Cell culture supernatant (300 μl) was obtained from cultured A549 cells and mixed with Apt63–SMB complex solution for 60 min.

Results and discussion

We designed capture probe Apt63, which consists of a biotin-terminated CD63 aptamer and a blocker that is partially complementary to CD63 aptamer.

When exosomes are present, the Apt63–SMB complex captures exosomes, leading to a conformational change of the CD63 aptamer that results in release of the blocker.

After removal of the remaining Apt63–SMB complexes and captured exosomes by a magnet, the blockers in the supernatant could be recognized by the CRISPR/Cas12a system and triggered trans-cleavage activity.

We successively optimized the concentration of Apt63 as a label on SMBs, the exosome-capture time, and the trans-cleavage time of CRISPR/Cas12a to achieve better exosome detection performance.

We first investigated the efficiency of exosome capture by SMBs linked to Apt63 in different concentrations by detecting the Cy3 label on the blocker.

Conclusions

We have developed a rapid and sensitive exosome detection method by integrating CD63 aptamer for exosome capture and CRISPR/Cas12a for signal amplification.

Acknowledgements

A machine generated summary based on the work of Zhao, Xianxian; Zhang, Wenqing; Qiu, Xiaopei; Mei, Qiang; Luo, Yang; Fu, Weiling, 2020 in Analytical and Bioanalytical Chemistry.

Titrating gene expression using libraries of systematically attenuated CRISPR guide RNAs.

https://doi.org/10.1038/s41587-019-0387-5

Abstract-Summary

We describe an approach to titrate expression of human genes using CRISPR interference and series of single-guide RNAs (sgRNAs) with systematically modulated activities.

These rules enabled us to synthesize a compact sgRNA library to titrate expression of ~ 2,400 genes essential for robust cell growth and to construct an in silico sgRNA library spanning the human genome.

Staging cells along a continuum of gene expression levels combined with single-cell RNA-seq readout revealed sharp transitions in cellular behaviors at gene-specific expression thresholds.

Main

Enabled by tools to titrate gene expression levels, such as series of promoters or hypomorphic mutants, the underlying expression–phenotype relationships have been explored systematically in yeast [221–223] and bacteria [224–227].

It is possible to titrate the expression of individual genes in mammalian systems by incorporating microRNA-binding sites of varied strength into the 3′-UTR of the endogenous locus [228] or using synthetic promoters and regulators [229], but these approaches require engineering of the endogenous locus for each target, limiting scalability and transferability across models.

We report a systematic approach to control DNA binding of dCas9 effectors through modified sgRNAs as a general method to titrate gene expression in mammalian cells.

As a starting point for analyses of expression–phenotype relationships in mammalian cells, we examined transcriptional phenotypes derived from single-cell RNA-seq at various expression levels of 25 essential genes.

Our data reveal gene-specific expression–phenotype relationships and expression-level-dependent cell responses at single-cell resolution, highlighting the utility of systematically attenuated sgRNAs in staging cells along a continuum of expression levels to explore fundamental biological questions.

Results

These results suggest that series of mismatched sgRNAs can be used to titrate gene expression at the single-cell level, but that mismatched sgRNA activity is modulated by complex factors.

These results suggest that systematically mismatched sgRNAs provide a general method to titrate CRISPRi activity and consequently, target gene expression.

We next sought to leverage our large-scale dataset of mismatched sgRNA activities to learn the underlying rules in a principled manner and enable predictions of intermediate-activity sgRNAs against other genes.

We selected 2,405 genes that we had found to be essential for robust growth of K562 cells in our large-scale screen, divided the relative activity space into six bins and attempted to select mismatched variants from each of the center four bins (relative activities between 0.1 and 0.9) for two sgRNA series targeting each gene.

Discussion

We describe the development of an approach to systematically titrate gene expression in human cells using allelic series of attenuated sgRNAs.

We highlight the utility of the approach by mapping gene expression levels to phenotypes with single-cell resolution, enabling identification of gene-specific viability thresholds and expression-level-dependent cell fates.

Our approach yields intermediate-activity sgRNAs in a predictable manner, is readily scalable to target any number of genes, in contrast to approaches that titrate gene expression using microRNAs or synthetic biology tools, and provides access to many expression levels of each gene in a single pooled experiment, in contrast to approaches that rely on small molecules to control (d)Cas9 activity.

These sgRNA series will also facilitate recapitulating gene expression levels of disease-relevant states, such as haploinsufficiency or partial loss-of-function, enabling efforts to identify suppressors or modifiers, or modeling quantitative trait loci associated with multigenic traits in conjunction with rich phenotyping to identify the mechanisms by which they interact and contribute to such traits.

Methods

To generate the list of targeting sgRNAs for the large-scale mismatched sgRNA library, hit genes from a growth screen performed in K562 cells with the CRISPRi v2 library [230] were selected by calculating a discriminant score (phenotype z score $\times - \log_{10}$(Mann–Whitney P)).

Relative activities were filtered for series in which the perfectly matched sgRNA had a growth phenotype greater than 5 z scores outside the distribution of negative control sgRNAs for all further analysis (3,147 and 2,029 sgRNA series for K562 and Jurkat cells, respectively).

Genes targeted by the compact allelic series library were required to have at least one perfectly matched sgRNA with a growth phenotype greater than two z scores outside the distribution of negative control sgRNAs ($\gamma < -0.04$) in a single replicate of a K562 pooled screen (this work or Horlbeck and others [230]).

Acknowledgements

A machine generated summary based on the work of Jost, Marco; Santos, Daniel A.; Saunders, Reuben A.; Horlbeck, Max A.; Hawkins, John S.; Scaria, Sonia M.; Norman, Thomas M.; Hussmann, Jeffrey A.; Liem, Christina R.; Gross, Carol A.; Weissman, Jonathan S, 2020 in Nature Biotechnology.

A CRISPR-Cas9-based reporter system for single-cell detection of extracellular vesicle-mediated functional transfer of RNA.

https://doi.org/10.1038/s41467-020-14977-8

Abstract-Summary

Whereas biological effects of EV-mediated RNA transfer are abundantly studied, regulatory pathways and mechanisms remain poorly defined due to a lack of suitable readout systems.

Using this CRISPR operated stoplight system for functional intercellular RNA exchange (CROSS-FIRE) we uncover various genes involved in EV subtype biogenesis that play a regulatory role in RNA transfer.

This approach allows the elucidation of regulatory mechanisms in EV-mediated RNA transfer at the level of EV biogenesis, endocytosis, intracellular trafficking, and RNA delivery.

Extended:

This modification further expands the potential of the CROSS-FIRE system, allowing future use of inducible or tissue-specific regulation of sgRNA expression in donor cells.

Introduction

A major drawback of mRNA-based systems is that it is inherently impossible to phenotypically distinguish between reporter activation as a result of the delivery of translated protein or the mRNA itself [231], which reduces the applicability of such systems to study RNA transfer specifically.

We establish protocols to study the effects of siRNA-mediated knockdown (KD) of single targets in both EV-acceptor and donor cells, as well as inhibitory compounds, on EV-mediated functional RNA transfer.

Using these protocols, we uncover several novel genes involved in the regulation of specific EV subtype biogenesis, as well as endocytosis and intracellular membrane trafficking that play a regulatory role in EV-mediated functional RNA delivery.

This novel approach allows the study of EV cargo processing in the context of functional RNA delivery, and may help to increase our understanding of the regulatory pathways that dictate the underlying processes.

Results

The observed low percentages of reporter activation do not necessarily reflect a low level of EV-mediated communication, but rather are the result of the low levels of sgRNA in EVs as we opted not to employ additional strategies for targeted loading of EVs with sgRNAs, such as RNA-binding proteins fused to EV-associated proteins, in order to study RNA loading and transfer in an unbiased manner.

These data show that the CROSS-FIRE reporter system is activated by EV-mediated sgRNA transfer.

We then employed this CROSS-FIRE based workflow to study EV-mediated RNA delivery and processing, by targeting various regulatory genes of endocytosis and intracellular membrane trafficking in HEK293T recipient reporter cells.

These data show that the CROSS-FIRE system provides a robust and scalable approach to study and uncover novel regulatory targets and pathways in intercellular RNA exchange in a direct co-culture setting, or using isolated EVs.

Discussion

Functional EV-mediated transfer of RNA molecules relies on uptake of the target cells, as well as subsequent specific intracellular trafficking and processing.

These results are in line with observations from Zomer and others using a Cre-LoxP-based reporter system to study EV-mediated cargo transfer from MDA-MB-231 donor cells to various reporter cell lines in co-culture experiments [232].

Based on these data it is thus tempting to conclude that tumor cells show higher levels of RNA transfer to other cells in general, as a result of increased EV secretion or more efficient uptake.

We optimized and demonstrate protocols that allow studying the role of single genetic targets in EV-mediated intercellular RNA transfer, by combining siRNA-mediated single target KD with both CROSS-FIRE co-culture and EV addition experiments.

A better understanding of these mechanisms may greatly aid in the design of EV-mediated RNA-delivery strategies, as EV uptake and EV cargo processing in EV acceptor cells strongly dictate efficiency of RNA delivery [233].

Methods

The MDA-MB-231 cells were maintained in the cell compartment in 15 ml serum-free OptiMEM, which was replaced every 48 h. For EV isolation, the serum-free conditioned medium was isolated from the concentrated cell compartment and cell debris was removed by 5 min centrifugation at $300 \times g$, followed by 15 min centrifugation at $2000 \times g$ ml.

Of EVs containing targeting sgRNAs or non-targeting sgRNAs, and for EV dose response addition experiments, HEK293T Stoplight[+]spCas9[+] reporter cells were cultured in 24-well plates in 1 ml culture medium, and EVs isolated from MDA-MB-231 cells expressing sgRNAs cultured in CELLine Adhere 1000 Bioreactors (Integra Biosciences) were added.

For EV addition experiments on siRNA-treated reporter cells, cells were plated in 96-well plate wells in a volume of 200 μl, and EVs isolated from MDA-MB-231 cells expressing sgRNAs cultured in T175 flasks were added every 24 h for six consecutive days.

Acknowledgements

A machine generated summary based on the work of de Jong, Olivier G.; Murphy, Daniel E.; Mäger, Imre; Willms, Eduard; Garcia-Guerra, Antonio; Gitz-Francois, Jerney J.; Lefferts, Juliet; Gupta, Dhanu; Steenbeek, Sander C.; van Rheenen, Jacco; El Andaloussi, Samir; Schiffelers, Raymond M.; Wood, Matthew J. A.; Vader, Pieter, 2020 in Nature Communications.

2.3 Potential Risks

Machine generated keywords: consequence, large deletion, deletion, cut, dsbs, casin-
duce, stem cell, sgrna, transient, genome engineering, normal, repair, cell cycle,
toxicity, introduction.

*CRISPR/Cas9 targeting events cause complex deletions and insertions at 17 sites in
the mouse genome.*

https://doi.org/10.1038/ncomms15464

Abstract-Summary

Although CRISPR/Cas9 genome editing has provided numerous opportunities to
interrogate the functional significance of any given genomic site, there is a paucity
of data on the extent of molecular scars inflicted on the mouse genome.

We interrogate the molecular consequences of CRISPR/Cas9-mediated deletions
at 17 sites in four loci of the mouse genome.

We sequence targeted sites in 632 founder mice and analyse 54 established lines.

Reliable deletion of juxtaposed sites is only achieved through two-step targeting.

Introduction

The application of the CRISPR/Cas9 system in genome engineering requires two
components, the single guide RNA (sgRNA) and the Cas9 nuclease [83].

Cas9 nuclease recognizes the protospacer adjacent motif (PAM) in the targeted
region, which is adjacent to sgRNA, and creates double-strand breaks.

CRISPR/Cas9 has been used successfully to disrupt individual and multiple
target genes and knock-in mice were generated to investigate biological functions
[124, 234–242].

Our laboratory has targeted 17 genomic sites in the mouse genome and the
respective mutations were analysed in 632 founders.

We investigated the molecular consequences on targeting sites with single sgRNAs
and identified prevalent asymmetric deletions, preferred sites at which deletion occurs
and large deletions.

Our analyses permitted an assessment of target specificities, deletion efficiencies
and size distributions based on one- and two-step targeting approaches.

Results

To obtain in-depth knowledge on the extent of molecular consequences at target sites,
we analysed deletions introduced at 17 loci in the mouse genome.

We initially failed to identify these large deletions due to the PCR screening
strategy, which typically amplifies short fragments ($\sim$400 bp) spanning individual
sites.

These large deletions were only detected using serial PCR primers spanning the
entire target region.

Our results indicate that although the simultaneous targeting can rapidly generate deletions of multiple sites, it frequently creates large deletions, possibly removing potential regulatory elements.

Although time consuming, two-step targeting appears to be the more reliable approach to precisely delete individual sites within a given locus.

Simultaneous targeting of sites separated by up to 23 kb results in the deletion of the entire region.

To restrict deletions to the desired sites we propose a sequential, two-step, targeting approach.

Discussion

Based on more than 630 founder mice and 54 established lines covering 17 genomic sites, we have now acquired a more detailed understanding of deletion patterns obtained on injection of single or multiple sgRNAs into mouse zygotes.

Preferential deletion patterns and large deletions were frequently obtained on targeting the mouse genome with single sgRNAs.

Our studies also highlight that attempts to individually delete juxtaposed sites in a given locus through the co-injection of several sgRNAs almost exclusively results in large deletions spanning the entire sequence between the outside sgRNAs.

We also obtained deletions of up to 24 kb induced by multiple sgRNAs targeting more than one juxtaposed site, which is in agreement with other studies [243].

Large deletions at target sites and even at some distance require additional analyses, possibly whole-genome sequencing, such as those reported to screen the off-target sites [244].

Methods

To target individual genomic sites, one or more than one sgRNA (up to three sgRNAs) were designed and injected independently or simultaneously.

To target more than one genomic site, several sgRNAs (up to a total of four sgRNAs, two sgRNAs for each site) were injected simultaneously or sequentially.

When we injected two sgRNAs, we normally used 50 ng μl^{-1} for each sgRNA.

The number of sgRNAs injected into the individual genomic sites: (1) Targeting individual genomic sites with single sgRNAs (Site B and C-1, two independent sgRNAs; site C-2/3, D-2, D-3, D-4(1), F, one sgRNA), (2) Targeting individual genomic sites with more than one sgRNA (Site A, C-1/3, D-1/2/3/4, E-1, two sgRNAs simultaneously; site D-4(2), E-2, E-3, E-4, E-5, three sgRNAs simultaneously) and (3) Targeting more than one genomic site with several sgRNAs (Site C-1/2, C-1/2/3, D-1/3/4, D1/2/3/4/5, total four sgRNAs simultaneously).

Additional information

How to cite this article: Shin, H. Y. and others CRISPR/Cas9 targeting events cause complex deletions and insertions at 17 sites in the mouse genome.

Publisher's note: Springer Nature remains neutral with regard to jurisdictional claims in published maps and institutional affiliations.

Acknowledgements

A machine generated summary based on the work of Shin, Ha Youn; Wang, Chaochen; Lee, Hye Kyung; Yoo, Kyung Hyun; Zeng, Xianke; Kuhns, Tyler; Yang, Chul Min; Mohr, Teresa; Liu, Chengyu; Hennighausen, Lothar, 2017 in Nature Communications.

p53 inhibits CRISPR–Cas9 engineering in human pluripotent stem cells.

https://doi.org/10.1038/s41591-018-0050-6

Abstract-Summary

Whereas some cell types are amenable to genome engineering, genomes of human pluripotent stem cells (hPSCs) have been difficult to engineer, with reduced efficiencies relative to tumour cell lines or mouse embryonic stem cells [119, 245–254].

This high efficiency of indel generation revealed that double-strand breaks (DSBs) induced by Cas9 are toxic and kill most hPSCs.

The toxic response to DSBs was P53/TP53-dependent, such that the efficiency of precise genome engineering in hPSCs with a wild-type P53 gene was severely reduced.

Our results indicate that Cas9 toxicity creates an obstacle to the high-throughput use of CRISPR/Cas9 for genome engineering and screening in hPSCs.

Main

To determine if this response is specific to hPSCs, we evaluated the non-targeting controls across pooled CRISPR screens in other cell lines.

The expression of P21 was increased between 3- and tenfold in the targeting sgRNAs compared to a non-targeting EGFP control sgRNA.

These results demonstrate that P53 is required for the toxic response to DSBs induced by Cas9.

P53 inhibition resulted in a 19-fold increase in successful insertions for 8402-iPSCs and a 16-fold increase for H1-hESCs, dramatically improving the efficiency of genome engineering in hPSCs.

We developed a highly efficient Cas9 system in hPSCs that is useful for screening and making engineered cells.

We found that DSBs induced by Cas9 triggered a P53-dependent toxic response that reduces the efficiency of engineering by at least an order of magnitude.

For ex vivo engineering, Cas9 toxicity combined with clonal expansion could potentially select for P53 mutant cells more tolerant of DNA damage.

Methods

05,873) to dissociate cell clumps to be replated in E8 plus thiazovivin (Selleckchem, S1459) at 0.2 μM. For lentiviral transduction, electroporation, pooled screening and live imaging of confluence, accutase (Gibco, A1110501) was used to create a single cell suspension, which was counted to accurately replate specific numbers of cells in E8 plus thiazovivin at 0.8 μM. Karyotyping was performed by Cell Line Genetics.

For OCT4 targeting assay, live cells were imaged for tdTomato fluorescence and then fixed, permeabilized, washed, incubated with peroxidase suppressor (Thermo) for 30 min, washed twice, and then blocked for 30 min (5% goat serum/0.1% Tween-20/PBS).

Dunnett's multiple comparisons test: Control (+) versus control (−), P = 0.0001 Control (+) versus mutant (−), P = 0.0001 Control (+) versus mutant (+), P = 0.0001 Statistical analysis of high-content imaging data quantified using Cell Profiler (version 2.1.1, revision 6c2d896) was conducted using PRISM software (version 7.0c).

Acknowledgements

A machine generated summary based on the work of Ihry, Robert J.; Worringer, Kathleen A.; Salick, Max R.; Frias, Elizabeth; Ho, Daniel; Theriault, Kraig; Kommineni, Sravya; Chen, Julie; Sondey, Marie; Ye, Chaoyang; Randhawa, Ranjit; Kulkarni, Tripti; Yang, Zinger; McAllister, Gregory; Russ, Carsten; Reece-Hoyes, John; Forrester, William; Hoffman, Gregory R.; Dolmetsch, Ricardo; Kaykas, Ajamete, 2018 in Nature Medicine.

CRISPR–Cas9 genome editing induces a p53-mediated DNA damage response.

https://doi.org/10.1038/s41591-018-0049-z

Abstract-Summary

We report that genome editing by CRISPR–Cas9 induces a p53-mediated DNA damage response and cell cycle arrest in immortalized human retinal pigment epithelial cells, leading to a selection against cells with a functional p53 pathway.

Main

To assess whether the transient Cas9 activity used in precision genome editing approaches would trigger a similar response, we transfected RPE1 p53$^{+/+}$ and p53$^{-/-}$ cells with ribonucleoprotein (RNP) complexes containing Cas9 and a guide.

p53 inhibition should increase the frequency of HDR and of precision genome editing, as it would permit cell cycle progression in the presence of Cas9-induced DSBs.

We cannot rule out that other mechanisms act downstream or parallel to p53 in limiting the efficiency of precision genome editing in normal human cells [255, 256].

We report here that genome editing by Cas9 in p53-proficient cells results in a DNA damage response, which causes a growth disadvantage/arrest, and decreases efficiency of precision genome editing.

Our results show that inhibiting DNA damage signaling can improve the efficiency of precision genome editing in normal, untransformed cells.

Methods

Cells were transduced with a lentivirus containing the plasmid pLenti-Cas9-Blast-sgHPRT, which contains both wild-type Cas9 and a guide sequence against the HPRT1 gene (GATGTGATGAAGGAGATGGG).

This screening method was used to verify the initial result of the p53 pathway knockdown seen in RPE1 cells after transcribed random sequence library screening [257].

The Brunello guide library (Addgene 73178), which targets 19,114 human genes with 4 guides each [257] was packaged in HEK293T cells as described above.

Populations of 80 million cells each of hTERT-RPE1 p53$^{+/+}$ (ATCC CRL-4000) and hTERT-RPE1 p53$^{-/-}$ were transduced with the Brunello guide library.

For cleaved caspase 3 and cell cycle analysis, we used an RNF2 locus guide, which seems to have no off-target effects [258] (crRNA sequence: GTCATCTTAGTCAT-TACCTG).

This screen was done in RPE1 p53$^{+/+}$ cells in two independent biological duplicates, with cell culture, sample preparation and next-generation sequencing occurring in different time points between the replicates.

Acknowledgements

A machine generated summary based on the work of Haapaniemi, Emma; Botla, Sandeep; Persson, Jenna; Schmierer, Bernhard; Taipale, Jussi, 2018 in Nature Medicine.

Repair of double-strand breaks induced by CRISPR–Cas9 leads to large deletions and complex rearrangements.

https://doi.org/10.1038/nbt.4192

Main

The vast majority of on-target DNA repair outcomes after Cas9 cutting in a variety of cell types are thought to be insertions and deletions (indels) of less than 20 bp [259–261].

We speculate that current assessments may have missed a substantial proportion of potential genotypes generated by on-target Cas9 cutting and repair, some of which may have potential pathogenic consequences following somatic editing of large populations of mitotically active cells.

We introduced Cas9 and gRNA constructs targeting intronic and exonic sites of PigA into JM8 mouse ES cells using PiggyBac transposition.

If intronic regulatory sequences were present around the exon, the DNA of cells sorted for retention of PigA expression would be wild type or contain small indels around the cut site.

To describe the genetic events underlying Cd9 loss, we isolated single-cell clones edited with the 3′ intronic guide, ascertained their expression status by flow cytometry and sequenced the area around the cut site using PacBio and Sanger technologies.

Methods

Cells were trypsinized, washed in M15, resuspended in M15 + LIF and seeded onto a gelatinized 24-well plate, containing the lipofectamine DNA complexes, at 3×10^5 cells per well.

Around 3×10^5 cells were collected on day 14 (or day 17, in case of the RPE1 cells), stained in PBS + 0.1% BSA for 30 min at room temperature with 1 μg/ml FLAER reagent (Cedarlane) or anti-Cd9-PE antibody (cat.

For single-cell cloning and PacBio experiments, cells were transfected in six-well plates with five times more cells and reagents, expanded onto 10-cm dishes and sorted by fluorescence-activated cell sorting for loss of FLAER or Cd9 staining on day 14 using MoFlow XDP (Beckman Coulter).

Genome coverage was calculated with "bedtools genomecov –dz" (v 2.27.1) using circular consensus sequences (PigA locus) or reconstructed alleles (Cd9 locus).

GFP-negative cells were sorted 4 d after the electroporation and plated into Methocult M3434 media (6,000 cells per 3 ml, StemCell Technologies).

Acknowledgements

A machine generated summary based on the work of Kosicki, Michael; Tomberg, Kärt; Bradley, Allan, 2018 in Nature Biotechnology.

Cas9 activates the p53 pathway and selects for p53-inactivating mutations.

https://doi.org/10.1038/s41588-020-0623-4

Abstract-Summary

Cas9 is commonly introduced into cell lines to enable CRISPR–Cas9-mediated genome editing.

Gene expression profiling of 165 pairs of human cancer cell lines and their Cas9-expressing derivatives revealed upregulation of the p53 pathway upon introduction of Cas9, specifically in wild-type TP53 (TP53-WT) cell lines.

Elevated levels of DNA repair were observed in Cas9-expressing cell lines.

Genetic characterization of 42 cell line pairs showed that introduction of Cas9 can lead to the emergence and expansion of p53-inactivating mutations.

Cas9 was less active in TP53-WT than in TP53-mutant cell lines, and Cas9-induced p53 pathway activation affected cellular sensitivity to both genetic and chemical perturbations.

Main

Immunoblotting confirmed p53 pathway activation upon Cas9 introduction into TP53-WT cells.

These findings demonstrate that Cas9-induced p53 activation cannot be explained by technical noise, by the effect of viral transduction, or by a general selection bottleneck.

Activation of the p53 pathway following Cas9 introduction suggests that p53 activity is a barrier that cells need to overcome to stably express Cas9.

To further test the functional implications of Cas9-induced p53 activation, we compared the response of parental and Cas9-expressing MCF7 cells to the MDM2 inhibitor nutlin-3.

Our findings suggest that Cas9-induced DNA damage may underlie p53 activation, but the molecular mechanisms that lead to this response, as well as those that mitigate it to allow Cas9 tolerance in the absence of genetic selection, remain to be elucidated.

Although we have not ruled out that some of the observed p53 activation could be attributed to the viral transduction itself [262, 263], and that the presence of a sgRNA could exacerbate p53 activation further [264–266], our findings demonstrate Cas9-specific p53 activation.

Methods

Meta Cas9 versus WT transcriptional signatures within each class of TP53 mutation statuses that were considered (TP53-WT or TP53-mutant, based on the functional mutation classification reported in Ref. [267]) were composed by taking the median value of the signal-to-noise ratio for each of the 10,174 genes across cell lines in the TP53 mutation class.

Dependency data were also subsctted to include only TP53-WT cell lines overlapping with available transcriptional data (n = 20 cell lines), and the same analysis of linear regression residuals was performed again, this time comparing cell lines whose transcriptional signature was positively enriched for the Hallmark p53 gene set versus cell lines without such enrichment, using a one-sided Wilcoxon rank test.

The significance of the differences in the transcriptional enrichment of the p53 and the DNA repair MSigDB Hallmark signatures between TP53-WT and TP53-mutant cell lines, and between Cas9 and empty or reporter vectors, were determined by a two-tailed Fisher's exact test.

Acknowledgements

A machine generated summary based on the work of Enache, Oana M.; Rendo, Veronica; Abdusamad, Mai; Lam, Daniel; Davison, Desiree; Pal, Sangita; Currimjee, Naomi; Hess, Julian; Pantel, Sasha; Nag, Anwesha; Thorner, Aaron R.; Doench, John G.; Vazquez, Francisca; Beroukhim, Rameen; Golub, Todd R.; Ben-David, Uri, 2020 in Nature Genetics.

References

1. Horvath, P., et al. 2008. Diversity, activity, and evolution of CRISPR loci in Streptococcus thermophilus. *Journal of Bacteriology* 190: 1401–1412.
2. Marraffini, L.A., and E.J. Sontheimer. 2010. CRISPR interference: RNA-directed adaptive immunity in bacteria and archaea. *Nature Rev. Genet.* 11: 181–190.

3. Marraffini, L.A., and E.J. Sontheimer. 2008. CRISPR interference limits horizontal gene transfer in staphylococci by targeting DNA. *Science* 322: 1843–1845.

4. Hale, C.R., et al. 2009. RNA-guided RNA cleavage by a CRISPR RNA-Cas protein complex. *Cell* 139: 945–956.

5. Lillestøl, R.K., et al. 2009. CRISPR families of the crenarchaeal genus Sulfolobus: Bidirectional transcription and dynamic properties. *Molecular Microbiology* 72: 259–272.

6. Pul, U. et al. Identification and characterization of E. coli CRISPR-cas promoters and their silencing by H-NS. Mol. Microbiol. 75, 1495–1512 (2010)

7. Barrangou, R., et al. 2007. CRISPR provides acquired resistance against viruses in prokaryotes. *Science* 315: 1709–1712.

8. Deveau, H., et al. 2008. Phage response to CRISPR-encoded resistance in Streptococcus thermophilus. *Journal of Bacteriology* 190: 1390–1400.

9. Girard, S.L., and S. Moineau. 2007. Analysis of two theta-replicating plasmids of Streptococcus thermophilus. *Plasmid* 58: 174–181.

10. Kiewiet, R., J. Kok, J.F. Seegers, G. Venema, and S. Bron. 1993. The mode of replication is a major factor in segregational plasmid instability in Lactococcus lactis. *Applied and Environment Microbiology* 59: 358–364.

11. Vaillancourt, K., et al. 2008. Role of galK and galM in galactose metabolism by Streptococcus thermophilus. *Applied and Environment Microbiology* 74: 1264–1267.

12. controlled expression of the green fluorescent protein. 2000. Nieto, C., Fernández de Palencia, P., López, P. & Espinosa, M. Construction of a tightly regulated plasmid vector for Streptococcus pneumoniae. *Plasmid* 43: 205–213.

13. Bolotin, A., B. Quinquis, A. Sorokin, and S.D. Ehrlich. 2005. Clustered regularly interspaced short palindrome repeats (CRISPRs) have spacers of extrachromosomal origin. *Microbiology* 151: 2551–2561.

14. Andersson, A.F., and J.F. Banfield. 2008. Virus population dynamics and acquired virus resistance in natural microbial communities. *Science* 320: 1047–1050.

15. Deveau, H., J.E. Garneau, and S. Moineau. 2010. CRISPR/Cas system and its role in phage-bacteria interactions. *Annual Review of Microbiology* 64: 475–493.

16. Horvath, P., and R. Barrangou. 2010. CRISPR/Cas, the immune system of bacteria and archaea. *Science* 327: 167–170.

17. Koonin, E. V. & Makarova, K. S. CRISPR-Cas: an adaptive immunity system in prokaryotes. F1000 Biol. Rep. 1, 95 (2009)

18. Sorek, R., V. Kunin, and P. Hugenholtz. 2008. CRISPR — a widespread system that provides acquired resistance against phages in bacteria and archaea. *Nature Rev. Microbiol.* 6: 181–186.

19. van der Oost, J., M.M. Jore, E.R. Westra, M. Lundgren, and S.J. Brouns. 2009. CRISPR-based adaptive and heritable immunity in prokaryotes. *Trends in Biochemical Sciences* 34: 401–407.

20. Mojica, F.J., C. Diez-Villasenor, J. Garcia-Martinez, and E. Soria. 2005. Intervening sequences of regularly spaced prokaryotic repeats derive from foreign genetic elements. *Journal of Molecular Evolution* 60: 174–182.

21. Pourcel, C., G. Salvignol, and G. Vergnaud. 2005. CRISPR elements in Yersinia pestis acquire new repeats by preferential uptake of bacteriophage DNA, and provide additional tools for evolutionary studies. *Microbiology* 151: 653–663.

22. van der Oost, J., and S.J. Brouns. 2009. RNAi: Prokaryotes get in on the act. *Cell* 139: 863–865.

23. Jansen, R., J.D. Embden, W. Gaastra, and L.M. Schouls. 2002. Identification of genes that are associated with DNA repeats in prokaryotes. *Molecular Microbiology* 43: 1565–1575.

24. Mojica, F.J., C. Ferrer, G. Juez, and F. Rodriguez-Valera. 1995. Long stretches of short tandem repeats are present in the largest replicons of the Archaea Haloferax mediterranei and Haloferax volcanii and could be involved in replicon partitioning. *Molecular Microbiology* 17: 85–93.

25. Nakata, A., M. Amemura, and K. Makino. 1989. Unusual nucleotide arrangement with repeated sequences in the Escherichia coli K-12 chromosome. *Journal of Bacteriology* 171: 3553–3556.

26. Waters, L.S., and G. Storz. 2009. Regulatory RNAs in bacteria. *Cell* 136: 615–628.
27. Makarova, K.S., L. Aravind, N.V. Grishin, I.B. Rogozin, and E.V.A. Koonin. 2002. DNA repair system specific for thermophilic Archaea and bacteria predicted by genomic context analysis. *Nucleic Acids Research* 30: 482–496.
28. Garneau, J.E., et al. 2010. The CRISPR/Cas bacterial immune system cleaves bacteriophage and plasmid DNA. *Nature* 468: 67–71.
29. Marraffini, L.A., and E.J. Sontheimer. 2010. Self versus non-self discrimination during CRISPR RNA-directed immunity. *Nature* 463: 568–571.
30. Drider, D., and C. Condon. 2004. The continuing story of endoribonuclease III. *Journal of Molecular Microbiology and Biotechnology* 8: 195–200.
31. Huntzinger, E., et al. 2005. Staphylococcus aureus RNAIII and the endoribonuclease III coordinately regulate spa gene expression. *EMBO Journal* 24: 824–835.
32. Nicholson, A.W. 1999. Function, mechanism and regulation of bacterial ribonucleases. *FEMS Microbiology Reviews* 23: 371–390.
33. Vogel, J., L. Argaman, E.G. Wagner, and S. Altuvia. 2004. The small RNA IstR inhibits synthesis of an SOS-induced toxic peptide. *Current Biology* 14: 2271–2276.
34. Opdyke, J.A., E.M. Fozo, M.R. Hemm, and G. Storz. 2010. RNase III participates in GadY-dependent cleavage of the gadX-gadW mRNA. *Journal of Molecular Biology* 406: 29–43.
35. Aliyari, R., and S.W. Ding. 2009. RNA-based viral immunity initiated by the Dicer family of host immune receptors. *Immunological Reviews* 227: 176–188.
36. Jinek, M., and J.A. Doudna. 2009. A three-dimensional view of the molecular machinery of RNA interference. *Nature* 457: 405–412.
37. Malone, C.D., and G.J. Hannon. 2009. Small RNAs as guardians of the genome. *Cell* 136: 656–668.
38. Meister, G., and T. Tuschl. 2004. Mechanisms of gene silencing by double-stranded RNA. *Nature* 431: 343–349.
39. Carmell, M.A., and G.J. Hannon. 2004. RNase III enzymes and the initiation of gene silencing. *Nature Struct. Mol. Biol.* 11: 214–218.
40. Condon, C. 2007. Maturation and degradation of RNA in bacteria. *Current Opinion in Microbiology* 10: 271–278.
41. Carte, J., R.Y. Wang, H. Li, R.M. Terns, and M.P. Terns. 2008. Cas6 is an endoribonuclease that generates guide RNAs for invader defense in prokaryotes. *Genes & Development* 22: 3489–3496.
42. Brouns, S.J., et al. 2008. Small CRISPR RNAs guide antiviral defense in prokaryotes. *Science* 321: 960–964.
43. Sharma, C.M., et al. 2010. The primary transcriptome of the major human pathogen Helicobacter pylori. *Nature* 464: 250–255.
44. Hille, F., et al. 2018. The biology of CRISPR-Cas: Backward and forward. *Cell* 172: 1239–1259.
45. Makarova, K.S., Y.I. Wolf, and E.V. Koonin. 2018. Classification and nomenclature of CRISPR-Cas systems: Where from here? *CRISPR J.* 1: 325–336.
46. Samai, P., et al. 2015. Co-transcriptional DNA and RNA cleavage during type III CRISPR-Cas immunity. *Cell* 161: 1164–1174.
47. Goldberg, G.W., W. Jiang, D. Bikard, and L.A. Marraffini. 2014. Conditional tolerance of temperate phages via transcription-dependent CRISPR–Cas targeting. *Nature* 514: 633–637.
48. Jackson, S.A., N. Birkholz, L.M. Malone, and P.C. Fineran. 2019. Imprecise spacer acquisition generates CRISPR-Cas immune diversity through primed adaptation. *Cell Host & Microbe* 25: 250–260.
49. Koonin, E.V., K.S. Makarova, and Y.I. Wolf. 2017. Evolutionary genomics of defense systems in archaea and bacteria. *Annual Review of Microbiology* 71: 233–261.
50. Bondy-Denomy, J., A. Pawluk, K.L. Maxwell, and A.R. Davidson. 2013. Bacteriophage genes that inactivate the CRISPR/Cas bacterial immune system. *Nature* 493: 429–432.
51. Pawluk, A., J. Bondy-Denomy, V.H.W. Cheung, K.L. Maxwell, and A.R. Davidson. 2014. A new group of phage anti-CRISPR genes inhibits the type I-E CRISPR–Cas system of Pseudomonas aeruginosa. *MBio* 5: e00896-e914.

52. Pawluk, A., et al. 2016. Inactivation of CRISPR–Cas systems by anti-CRISPR proteins in diverse bacterial species. *Nature Microbiology* 1: 16085.
53. van Belkum, A., et al. 2015. Phylogenetic distribution of CRISPR–Cas systems in antibiotic-resistant Pseudomonas aeruginosa. *MBio* 6: e01796-e1815.
54. Makarova, K.S., et al. 2015. An updated evolutionary classification of CRISPR–Cas systems. *Nature Reviews Microbiology* 13: 722–736.
55. Marino, N.D., et al. 2018. Discovery of widespread type I and type V CRISPR–Cas inhibitors. *Science* 362: 240–242.
56. Bryson, A. L. et al. Covalent modification of bacteriophage T4 DNA inhibits CRISPR–Cas9. mBio 6, e00648–15 (2015).
57. Strotskaya, A., et al. 2017. The action of Escherichia coli CRISPR–Cas system on lytic bacteriophages with different lifestyles and development strategies. *Nucleic Acids Research* 45: 1946–1957.
58. Vlot, M., et al. 2018. Bacteriophage DNA glucosylation impairs target DNA binding by type I and II but not by type V CRISPR–Cas effector complexes. *Nucleic Acids Research* 46: 873–885.
59. Huang, L.H., C.M. Farnet, K.C. Ehrlich, and M. Ehrlich. 1982. Digestion of highly modified bacteriophage DNA by restriction endonucleases. *Nucleic Acids Research* 10: 1579–1591.
60. Chaikeeratisak, V., et al. 2017. Assembly of a nucleus-like structure during viral replication in bacteria. *Science* 355: 194–197.
61. Chaikeeratisak, V., et al. 2017. The phage nucleus and tubulin spindle are conserved among large Pseudomonas phages. *Cell Reports* 20: 1563–1571.
62. Abudayyeh, O. O. et al. C2c2 is a single-component programmable RNA-guided RNA-targeting CRISPR effector. Science 353, aaf5573 (2016).
63. Gootenberg, J.S., et al. 2017. Nucleic acid detection with CRISPR-Cas13a/C2c2. *Science* 356: 438–442.
64. Tamulaitis, G., et al. 2014. Programmable RNA shredding by the type III-A CRISPR-Cas system of Streptococcus thermophilus. *Molecular Cell* 56: 506–517.
65. Kazlauskiene, M., G. Kostiuk, Č Venclovas, G. Tamulaitis, and V. Siksnys. 2017. A cyclic oligonucleotide signaling pathway in type III CRISPR-Cas systems. *Science* 357: 605–609.
66. Niewoehner, O., et al. 2017. Type III CRISPR-Cas systems produce cyclic oligoadenylate second messengers. *Nature* 548: 543–548.
67. Rouillon, C., Athukoralage, J. S., Graham, S., Grüschow, S. & White, M. F. Control of cyclic oligoadenylate synthesis in a type III CRISPR system. eLife 7, e36734 (2018).
68. Makarova, K.S., V. Anantharaman, N.V. Grishin, E.V. Koonin, and L. Aravind. 2014. CARF and WYL domains: Ligand-binding regulators of prokaryotic defense systems. *Frontiers in Genetics* 5: 102.
69. Rostøl, J.T., and L.A. Marraffini. 2019. Non-specific degradation of transcripts promotes plasmid clearance during type III-A CRISPR-Cas immunity. *Nature Microbiology* 4: 656–662.
70. Pyenson, N.C., K. Gayvert, A. Varble, O. Elemento, and L.A. Marraffini. 2017. Broad targeting specificity during bacterial type III CRISPR-Cas immunity constrains viral escape. *Cell Host & Microbe* 22: 343–353.
71. Deng, L., R.A. Garrett, S.A. Shah, X. Peng, and Q. She. 2013. A novel interference mechanism by a type IIIB CRISPR-Cmr module in Sulfolobus. *Molecular Microbiology* 87: 1088–1099.
72. Jiang, W., P. Samai, and L.A. Marraffini. 2016. Degradation of phage transcripts by CRISPR-associated RNases enables type III CRISPR-Cas immunity. *Cell* 164: 710–721.
73. Athukoralage, J.S., C. Rouillon, S. Graham, S. Grüschow, and M.F. White. 2018. Ring nucleases deactivate type III CRISPR ribonucleases by degrading cyclic oligoadenylate. *Nature* 562: 277–280.
74. Borges, A.L., A.R. Davidson, and J. Bondy-Denomy. 2017. The discovery, mechanisms, and evolutionary impact of anti-CRISPRs. *Annu. Rev. Virol.* 4: 37–59.
75. Hwang, S., and K.L. Maxwell. 2019. Meet the anti-CRISPRs: Widespread protein inhibitors of CRISPR-Cas systems. *CRISPR J.* 2: 23–30.

76. Bhoobalan-Chitty, Y., T.B. Johansen, N. Di Cianni, and X. Peng. 2019. Inhibition of type III CRISPR-Cas immunity by an archaeal virus-encoded anti-CRISPR protein. *Cell* 179: 448–458.

77. Keller, J., et al. 2007. Crystal structure of AFV3-109, a highly conserved protein from crenarchaeal viruses. *Virol. J.* 4: 12.

78. Whiteley, A.T., et al. 2019. Bacterial cGAS-like enzymes synthesize diverse nucleotide signals. *Nature* 567: 194–199.

79. Maelfait, J., and J. Rehwinkel. 2017. RECONsidering sensing of cyclic dinucleotides. *Immunity* 46: 337–339.

80. Cohen, D., et al. 2019. Cyclic GMP-AMP signalling protects bacteria against viral infection. *Nature* 574: 691–695.

81. Wiedenheft, B., S.H. Sternberg, and J.A. Doudna. 2012. RNA-guided genetic silencing systems in bacteria and archaea. *Nature* 482: 331–338.

82. Deltcheva, E., et al. 2011. CRISPR RNA maturation by trans-encoded small RNA and host factor RNase III. *Nature* 471: 602–607.

83. Jinek, M., et al. 2012. A programmable dual-RNA-guided DNA endonuclease in adaptive bacterial immunity. *Science* 337: 816–821.

84. Gasiunas, G., R. Barrangou, P. Horvath, and V. Siksnys. 2012. Cas9-crRNA ribonucleoprotein complex mediates specific DNA cleavage for adaptive immunity in bacteria. *Proceedings of the National academy of Sciences of the United States of America* 109: E2579–E2586.

85. von Hippel, P.H., and O.G. Berg. 1989. Facilitated target location in biological systems. *Journal of Biological Chemistry* 264: 675–678.

86. Gorman, J., A.J. Plys, M.-L. Visnapuu, E. Alani, and E.C. Greene. 2010. Visualizing one-dimensional diffusion of eukaryotic DNA repair factors along a chromatin lattice. *Nature Struct. Mol. Biol.* 17: 932–938.

87. Wang, F., et al. 2013. The promoter-search mechanism of Escherichia coli RNA polymerase is dominated by three-dimensional diffusion. *Nature Struct. Mol. Biol.* 20: 174–181.

88. Gorman, J., et al. 2012. Single-molecule imaging reveals target-search mechanisms during DNA mismatch repair. *Proceedings of the National academy of Sciences of the United States of America* 109: E3074–E3083.

89. Sashital, D.G., B. Wiedenheft, and J.A. Doudna. 2012. Mechanism of foreign DNA selection in a bacterial adaptive immune system. *Molecular Cell* 46: 606–615.

90. Jiang, W., D. Bikard, D. Cox, F. Zhang, and L.A. Marraffini. 2013. RNA-guided editing of bacterial genomes using CRISPR-Cas systems. *Nature Biotechnol.* 31: 233–239.

91. Pattanayak, V., et al. 2013. High-throughput profiling of off-target DNA cleavage reveals RNA-programmed Cas9 nuclease specificity. *Nature Biotechnol.* 31: 839–843.

92. Hsu, P.D., et al. 2013. DNA targeting specificity of RNA-guided Cas9 nucleases. *Nature Biotechnol.* 31: 827–832.

93. Fu, Y., et al. 2013. High-frequency off-target mutagenesis induced by CRISPR-Cas nucleases in human cells. *Nature Biotechnol.* 31: 822–826.

94. McGraw, C.M., R.C. Samaco, and H.Y. Zoghbi. 2011. Adult neural function requires MeCP2. *Science* 333: 186.

95. Shalem, O., et al. 2014. Genome-scale CRISPR-Cas9 knockout screening in human cells. *Science* 343: 84–87.

96. Wang, T., J.J. Wei, D.M. Sabatini, and E.S. Lander. 2014. Genetic screens in human cells using the CRISPR-Cas9 system. *Science* 343: 80–84.

97. Maeder, M.L., et al. 2013. CRISPR RNA-guided activation of endogenous human genes. *Nature Methods* 10: 977–979.

98. Perez-Pinera, P., et al. 2013. RNA-guided gene activation by CRISPR-Cas9-based transcription factors. *Nature Methods* 10: 973–976.

99. Mali, P., et al. 2013. CAS9 transcriptional activators for target specificity screening and paired nickases for cooperative genome engineering. *Nature Biotechnol.* 31: 833–838.

100. Konermann, S., et al. 2013. Optical control of mammalian endogenous transcription and epigenetic states. *Nature* 500: 472–476.

101. Prahallad, A., et al. 2012. Unresponsiveness of colon cancer to BRAF(V600E) inhibition through feedback activation of EGFR. *Nature* 483: 100–103.
102. Corcoran, R.B., et al. 2012. EGFR-mediated re-activation of MAPK signaling contributes to insensitivity of BRAF mutant colorectal cancers to RAF inhibition with vemurafenib. *Cancer Discovery* 2: 227–235.
103. Johannessen, C.M., et al. 2013. A melanocyte lineage program confers resistance to MAP kinase pathway inhibition. *Nature* 504: 138–142.
104. Barretina, J., et al. 2012. The Cancer Cell Line Encyclopedia enables predictive modelling of anticancer drug sensitivity. *Nature* 483: 603–607.
105. Lin, W.M., et al. 2008. Modeling genomic diversity and tumor dependency in malignant melanoma. *Cancer Research* 68: 664–673.
106. Chang, N., et al. 2013. Genome editing with RNA-guided Cas9 nuclease in zebrafish embryos. *Cell Research* 23: 465–472.
107. Cho, S.W., S. Kim, J.M. Kim, and J.S. Kim. 2013. Targeted genome engineering in human cells with the Cas9 RNA-guided endonuclease. *Nature Biotechnology* 31: 230–232.
108. Cong, L., et al. 2013. Multiplex genome engineering using CRISPR/Cas systems. *Science* 339: 819–823.
109. Friedland, A.E. et al. Heritable genome editing in C. elegans via a CRISPR-Cas9 system. Nat. Methods 10, 741–743 (2013).
110. Hsu, P.D., E.S. Lander, and F. Zhang. 2014. Development and applications of CRISPR–Cas9 for genome engineering. *Cell* 157: 1262–1278.
111. Hwang, W.Y., et al. 2013. Efficient genome editing in zebrafish using a CRISPR-Cas system. *Nature Biotechnology* 31: 227–229.
112. Ikmi, A., S.A. McKinney, K.M. Delventhal, and M.C. Gibson. 2014. TALEN and CRISPR/Cas9-mediated genome editing in the early-branching metazoan Nematostella vectensis. *Nature Communications* 5: 5486.
113. Irion, U., J. Krauss, and C. Nusslein-Volhard. 2014. Precise and efficient genome editing in zebrafish using the CRISPR/Cas9 system. *Development* 141 (24): 4827–4830.
114. Jinek, M. et al. RNA-programmed genome editing in human cells. Elife 2, e00471 (2013).
115. Li, D., et al. 2013. Heritable gene targeting in the mouse and rat using a CRISPR-Cas system. *Nature Biotechnology* 31: 681–683.
116. Li, W., F. Teng, T. Li, and Q. Zhou. 2013. Simultaneous generation and germline transmission of multiple gene mutations in rat using CRISPR-Cas systems. *Nature Biotechnology* 31: 684–686.
117. Long, C., J.R. McAnally, J.M. Shelton, A.A. Mireault, R. Bassel-Duby, and E.N. Olson. 2014. Prevention of muscular dystrophy in mice by CRISPR/Cas9-mediated editing of germline DNA. *Science* 345: 1184–1188.
118. Ma, Y., X. Zhang, B. Shen, Y. Lu, W. Chen, J. Ma, L. Bai, X. Huang, and L. Zhang. 2014. Generating rats with conditional alleles using CRISPR/Cas9. *Cell Research* 24: 122–125.
119. Mali, P., et al. 2013. RNA-guided human genome engineering via Cas9. *Science* 339: 823–826.
120. Niu, Y., B. Shen, Y. Cui, Y. Chen, J. Wang, L. Wang, Y. Kang, X. Zhao, W. Si, W. Li, et al. 2014. Generation of gene-modified cynomolgus monkey via Cas9/RNA-mediated gene targeting in one-cell embryos. *Cell* 156: 836–843.
121. Smith, C., L. Abalde-Atristain, C. He, B.R. Brodsky, E.M. Braunstein, P. Chaudhari, Y.Y. Jang, L. Cheng, and Z. Ye. 2014. Efficient and allele-specific genome editing of disease loci in human iPSCs. *Molecular Therapy* 23: 570–577.
122. Wu, Y., D. Liang, Y. Wang, M. Bai, W. Tang, S. Bao, Z. Yan, D. Li, and J. Li. 2013. Correction of a genetic disease in mouse via use of CRISPR-Cas9. *Cell Stem Cell* 13: 659–662.
123. Wu, Y., H. Zhou, X. Fan, Y. Zhang, M. Zhang, Y. Wang, Z. Xie, M. Bai, Q. Yin, D. Liang, et al. 2014. Correction of a genetic disease by CRISPR-Cas9-mediated gene editing in mouse spermatogonial stem cells. *Cell Research* 25: 67–79.
124. Yang, H., H. Wang, C.S. Shivalila, A.W. Cheng, L. Shi, and R. Jaenisch. 2013. One-step generation of mice carrying reporter and conditional alleles by CRISPR/Cas-mediated genome engineering. *Cell* 154: 1370–1379.

125. Smith, C., A. Gore, W. Yan, L. Abalde-Atristain, Z. Li, C. He, Y. Wang, R.A. Brodsky, K. Zhang, L. Cheng, et al. 2014. Whole-genome sequencing analysis reveals high specificity of CRISPR/Cas9 and TALEN-Based Genome Editing in Human iPSCs. *Cell Stem Cell* 15: 12–13.

126. Suzuki, K., C. Yu, J. Qu, M. Li, X. Yao, T. Yuan, A. Goebl, S. Tang, R. Ren, E. Aizawa, et al. 2014. Targeted gene correction minimally impacts whole-genome mutational load in human-disease-specific induced pluripotent stem cell clones. *Cell Stem Cell* 15: 31–36.

127. Veres, A., B.S. Gosis, Q. Ding, R. Collins, A. Ragavendran, H. Brand, S. Erdin, M.E. Talkowski, and K. Musunuru. 2014. Low incidence of off-target mutations in individual CRISPR-Cas9 and TALEN targeted human stem cell clones detected by whole-genome sequencing. *Cell Stem Cell* 15: 27–30.

128. Cradick, T.J., E.J. Fine, C.J. Antico, and G. Bao. 2013. CRISPR/Cas9 systems targeting beta-globin and CCR5 genes have substantial off-target activity. *Nucleic Acids Research* 41: 9584–9592.

129. Dean, F.B., S. Hosono, L. Fang, X. Wu, A.F. Faruqi, P. Bray-Ward, Z. Sun, Q. Zong, Y. Du, J. Du, et al. 2002. Comprehensive human genome amplification using multiple displacement amplification. *Proceedings of the National academy of Sciences of the United States of America* 99: 5261–5266.

130. Hosono, S., A.F. Faruqi, F.B. Dean, Y. Du, Z. Sun, X. Wu, J. Du, S.F. Kingsmore, M. Egholm, and R.S. Lasken. 2003. Unbiased whole-genome amplification directly from clinical samples. *Genome Research* 13: 954–964.

131. San Filippo, J., P. Sung, and H. Klein. 2008. Mechanism of eukaryotic homologous recombination. *Annual Review of Biochemistry* 77: 229–257.

132. Baltimore, B.D., P. Berg, M. Botchan, D. Carroll, R.A. Charo, G. Church, J.E. Corn, G.Q. Daley, J.A. Doudna, M. Fenner, et al. 2015. A prudent path forward for genomic engineering and germline gene modification. *Science* 348: 36–38.

133. Cyranoski, D. 2015. Ethics of embryo editing divides scientists. *Nature* 519: 272.

134. Lanphier, E., F. Urnov, S.E. Haecker, M. Werner, and J. Smolenski. 2015. Don't edit the human germ line. *Nature* 519: 410–411.

135. Bansal, V., and O. Libiger. 2011. A probabilistic method for the detection and genotyping of small indels from population-scale sequence data. *Bioinformatics* 27: 2047–2053.

136. Papathanos, P.A., N. Windbichler, M. Menichelli, A. Burt, and A. Crisanti. 2009. The vasa regulatory region mediates germline expression and maternal transmission of proteins in the malaria mosquito Anopheles gambiae: A versatile tool for genetic control strategies. *BMC Molecular Biology* 10: 65.

137. Konet, D.S., et al. 2007. Short-hairpin RNA expressed from polymerase III promoters mediates RNA interference in mosquito cells. *Insect Molecular Biology* 16: 199–206.

138. Liang, P., Y. Xu, X. Zhang, C. Ding, R. Huang, Z. Zhang, J. Lv, X. Xie, Y. Chen, Y. Li, Y. Sun, Y. Bai, Z. Songyang, W. Ma, C. Zhou, and J. Huang. 2015. CRISPR/Cas9-mediated gene editing in human tripronuclear zygotes. *Protein & Cell* 6: 363–372.

139. Kang, X., W. He, Y. Huang, Q. Yu, Y. Chen, X. Gao, X. Sun, and Y. Fan. 2016. Introducing precise genetic modifications into human 3PN embryos by CRISPR/Cas-mediated genome editing. *Journal of Assisted Reproduction and Genetics* 33: 581–588.

140. Shen, B., et al. 2013. Generation of gene-modified mice via Cas9/RNA-mediated gene targeting. *Cell Research* 23: 720–723.

141. Miura, H., et al. 2015. CRISPR/Cas9-based generation of knockdown mice by intronic insertion of artificial microRNA using longer single-stranded DNA. *Science and Reports* 5: 12799.

142. Bishop KA, et al. CRISPR/Cas9-mediated insertion of loxP sites in the mouse Dock7 gene provides an effective alternative to use of targeted embryonic stem cells. G3 Bethesda Md. 2016;6:2051–61.

143. Jacobi, A.M., et al. 2017. Simplified CRISPR tools for efficient genome editing and streamlined protocols for their delivery into mammalian cells and mouse zygotes. *Methods*. https://doi.org/10.1016/j.ymeth.2017.03.021.

144. Frum, T., et al. 2013. Oct4 cell-autonomously promotes primitive endoderm development in the mouse blastocyst. *Developmental Cell* 25: 610–622.
145. Macaulay, I.C., et al. 2015. G&T-seq: Parallel sequencing of single-cell genomes and transcriptomes. *Nature Methods* 12: 519–522.
146. Yan, L., et al. 2013. Single-cell RNA-seq profiling of human preimplantation embryos and embryonic stem cells. *Nature Structural & Molecular Biology* 20: 1131–1139.
147. Joung, J., et al. 2017. Genome-scale CRISPR–Cas9 knockout and transcriptional activation screening. *Nat. Protocols* 12: 828–863.
148. Trapnell, C., et al. 2014. The dynamics and regulators of cell fate decisions are revealed by pseudotemporal ordering of single cells. *Nature Biotechnology* 32: 381–386.
149. Setty, M., et al. 2016. Wishbone identifies bifurcating developmental trajectories from single-cell data. *Nature Biotechnology* 34: 637–645.
150. Rizvi, A.H., et al. 2017. Single-cell topological RNA-seq analysis reveals insights into cellular differentiation and development. *Nature Biotechnology* 35: 551–560.
151. Shin, J., et al. 2015. Single-cell RNA-Seq with waterfall reveals molecular cascades underlying adult neurogenesis. *Cell Stem Cell* 17: 360–372.
152. Furchtgott, L.A., Melton, S., Menon, V. & Ramanathan, S. Discovering sparse transcription factor codes for cell states and state transitions during development. eLife 6, e20488 (2017).
153. Kretzschmar, K., and F.M. Watt. 2012. Lineage tracing. *Cell* 148: 33–45.
154. Woodworth, M.B., K.M. Girskis, and C.A. Walsh. 2017. Building a lineage from single cells: Genetic techniques for cell lineage tracking. *Nature Reviews Genetics* 18: 230–244.
155. Howe, D.G., et al. 2013. ZFIN, the Zebrafish Model Organism Database: Increased support for mutants and transgenics. *Nucleic Acids Research* 41: D854–D860.
156. Wilson, S.W., M. Brand, and J.S. Eisen. 2002. Patterning the zebrafish central nervous system. *Results and Problems in Cell Differentiation* 40: 181–215.
157. McKenna, A. et al. Whole-organism lineage tracing by combinatorial and cumulative genome editing. Science 353, aaf7907 (2016).
158. Grandel, H., J. Kaslin, J. Ganz, I. Wenzel, and M. Brand. 2006. Neural stem cells and neurogenesis in the adult zebrafish brain: Origin, proliferation dynamics, migration and cell fate. *Developmental Biology* 295: 263–277.
159. Junker, J.P., et al. 2017. Massively parallel clonal analysis using CRISPR/Cas9 induced genetic scars. *Preprint at bioRxiv.* https://doi.org/10.1101/056499.
160. Schmidt, S.T., S.M. Zimmerman, J. Wang, S.K. Kim, and S.R. Quake. 2017. Quantitative analysis of synthetic cell lineage tracing using nuclease barcoding. *ACS Synthetic Biology* 6: 936–942.
161. Klein, A.M., et al. 2015. Droplet barcoding for single-cell transcriptomics applied to embryonic stem cells. *Cell* 161: 1187–1201.
162. Zilionis, R., et al. 2017. Single-cell barcoding and sequencing using droplet microfluidics. *Nature Protocols* 12: 44–73.
163. Alemany, A., et al. 2018. Whole-organism clone tracing using single-cell sequencing. *Nature.* https://doi.org/10.1038/nature25969.
164. Raj, B., et al. 2018. Simultaneous single-cell profiling of lineages and cell types in the vertebrate brain. *Nature Biotechnology.* https://doi.org/10.1038/nbt.4103.
165. Gantz, V.M., et al. 2015. Highly efficient Cas9-mediated gene drive for population modification of the malaria vector mosquito Anopheles stephensi. *Proceedings of the National academy of Sciences of the United States of America* 112: E6736–E6743.
166. Hammond, A., et al. 2016. A CRISPR-Cas9 gene drive system targeting female reproduction in the malaria mosquito vector Anopheles gambiae. *Nature Biotechnology* 34: 78–83.
167. Burt, A. 2003. Site-specific selfish genes as tools for the control and genetic engineering of natural populations. *Proceedings of the Royal Society of London, Series B: Biological Sciences* 270: 921–928.
168. Deredec, A., H.C. Godfray, and A. Burt. 2011. Requirements for effective malaria control with homing endonuclease genes. *Proceedings of the National academy of Sciences of the United States of America* 108: E874–E880.

169. Hamilton, W.D. Extraordinary sex ratios. A sex-ratio theory for sex linkage and inbreeding has new implications in cytogenetics and entomology. Science 156, 477–488 (1967).

170. Galizi, R., et al. 2014. A synthetic sex ratio distortion system for the control of the human malaria mosquito. *Nature Communications* 5: 3977.

171. Murray, S.M., S.Y. Yang, and M. Van Doren. 2010. Germ cell sex determination: A collaboration between soma and germline. *Current Opinion in Cell Biology* 22: 722–729.

172. Curtis, C.F. 1968. Possible use of translocations to fix desirable genes in insect pest populations. *Nature* 218: 368–369.

173. KaramiNejadRanjbar, M., et al. 2018. Consequences of resistance evolution in a Cas9-based sex conversion-suppression gene drive for insect pest management. *Proceedings of the National academy of Sciences of the United States of America* 115: 6189–6194.

174. National Academies of Sciences, Engineering & Medicine. Gene Drives on the Horizon: Advancing Science, Navigating Uncertainty, and Aligning Research with Public Values (National Academies Press, Washington, DC, 2016).

175. Dominguez, A.A., W.A. Lim, and L.S. Qi. 2016. Beyond editing: Repurposing CRISPR–Cas9 for precision genome regulation and interrogation. *Nature Reviews Molecular Cell Biology* 17: 5–15.

176. Komor, A.C., A.H. Badran, and D.R. Liu. 2017. CRISPR-based technologies for the manipulation of eukaryotic genomes. *Cell* 168: 20–36.

177. Nielsen, A.A., and C.A. Voigt. 2014. Multi-input CRISPR/Cas genetic circuits that interface host regulatory networks. *Molecular Systems Biology* 10: 763–763.

178. Gander, M.W., J.D. Vrana, W.E. Voje, J.M. Carothers, and E. Klavins. 2017. Digital logic circuits in yeast with CRISPR-dCas9 NOR gates. *Nature Communications* 8: 15459.

179. Pawluk, A., et al. 2016. Naturally occurring off-switches for CRISPR–Cas9. *Cell* 167: 1829–1838.

180. Rauch, B.J., et al. 2017. Inhibition of CRISPR–Cas9 with bacteriophage proteins. *Cell* 168: 150–158.

181. Harrington, L.B., et al. 2017. A broad-spectrum inhibitor of CRISPR–Cas9. *Cell* 170: 1224–1233.e15.

182. Li, J., Xu, Z., Chupalov, A. & Marchisio, M. A. Anti-CRISPR-based biosensors in the yeast S. cerevisiae. J. Biol. Eng. 12, 11 (2018).

183. Basgall, E.M., et al. 2018. Gene drive inhibition by the anti-CRISPR proteins AcrIIA2 and AcrIIA4 in Saccharomyces cerevisiae. *Microbiology* 164: 464–474.

184. Liu, X.S., et al. 2018. Rescue of fragile X syndrome neurons by DNA methylation editing of the FMR1 gene. *Cell* 172: 979-992.e6.

185. Gao, Y., et al. 2016. Complex transcriptional modulation with orthogonal and inducible dCas9 regulators. *Nature Methods* 13: 1043–1049.

186. Kipniss, N.H., et al. 2017. Engineering cell sensing and responses using a GPCR-coupled CRISPR-Cas system. *Nature Communications* 8: 2212.

187. Gantz, V. M. & Bier, E. Genome editing. The mutagenic chain reaction: a method for converting heterozygous to homozygous mutations. Science 348, 442–444 (2015).

188. Kyrou, K., et al. 2018. A CRISPR–Cas9 gene drive targeting doublesex causes complete population suppression in caged Anopheles gambiae mosquitoes. *Nature Biotechnology* 36: 1062–1066.

189. Burstein, D., et al. 2017. New CRISPR–Cas systems from uncultivated microbes. *Nature* 542: 237–241.

190. Koonin, E.V., K.S. Makarova, and F. Zhang. 2017. Diversity, classification and evolution of CRISPR-Cas systems. *Current Opinion in Microbiology* 37: 67–78.

191. Yamano, T., et al. 2016. Crystal structure of Cpf1 in complex with guide RNA and target DNA. *Cell* 165: 949–962.

192. Yang, H., P. Gao, K.R. Rajashankar, and D.J. Patel. 2016. PAM-dependent target DNA recognition and cleavage by C2c1 CRISPR–Cas endonuclease. *Cell* 167: 1814-1828.e1812.

193. Zetsche, B., et al. 2015. Cpf1 is a single RNA-guided endonuclease of a class 2 CRISPR-Cas system. *Cell* 163: 759–771.

194. Oakes, B.L., D.C. Nadler, and D.F. Savage. 2014. Protein engineering of Cas9 for enhanced function. *Methods in Enzymology* 546: 491–511.
195. Oakes, B.L., et al. 2016. Profiling of engineering hotspots identifies an allosteric CRISPR-Cas9 switch. *Nature Biotechnology* 34: 646–651.
196. Yamano, T., et al. 2017. Structural basis for the canonical and non-canonical PAM recognition by CRISPR–Cpf1. *Molecular Cell* 67: 633-645.e633.
197. Jaitin, D.A., et al. 2016. *Cell* 167: 1883-1896.e15.
198. Adamson, B., et al. 2016. A multiplexed single-cell CRISPR screening platform enables systematic dissection of the unfolded protein response. *Cell* 167: 1867-1882.e21.
199. Dixit, A., et al. 2016. *Cell* 167: 1853-1866.e17.
200. Datlinger, P., et al. 2017. *Nature Methods* 14: 297–301.
201. Stoeckius, M., et al. 2017. *Nature Methods* 14: 865–868.
202. Peterson, V.M., et al. 2017. *Nature Biotechnology* 35: 936–939.
203. Stoeckius, M., et al. 2018. *Genome Biology* 19: 224.
204. Koonin, E.V., and K.S. Makarova. 2017. Mobile genetic elements and evolution of CRISPR-Cas systems: All the way there and back. *Genome Biology and Evolution* 9: 2812–2825.
205. Broecker, F., and K. Moelling. 2019. Evolution of immune systems from viruses and transposable elements. *Frontiers in Microbiology* 10: 51.
206. Kapitonov, V.V., K.S. Makarova, and E.V. Koonin. 2016. ISC, a novel group of bacterial and archaeal DNA transposons that encode Cas9 homologs. *Journal of Bacteriology* 198: 797–807.
207. Shmakov, S., et al. 2015. Discovery and functional characterization of diverse class 2 CRISPR-Cas systems. *Molecular Cell* 60: 385–397.
208. Krupovic, M., P. Béguin, and E.V. Koonin. 2017. Casposons: Mobile genetic elements that gave rise to the CRISPR-Cas adaptation machinery. *Current Opinion in Microbiology* 38: 36–43.
209. Peters, J.E., K.S. Makarova, S. Shmakov, and E.V. Koonin. 2017. Recruitment of CRISPR-Cas systems by Tn7-like transposons. *Proceedings of the National academy of Sciences of the United States of America* 114: E7358–E7366.
210. Peters, J. E. Tn7. Microbiol. Spectr. 2, MDNA3–0010–2014 (2014).
211. Choi, K.Y., J.M. Spencer, and N.L. Craig. 2014. The Tn7 transposition regulator TnsC interacts with the transposase subunit TnsB and target selector TnsD. *Proceedings of the National academy of Sciences of the United States of America* 111: E2858–E2865.
212. Wiedenheft, B., et al. 2011. RNA-guided complex from a bacterial immune system enhances target recognition through seed sequence interactions. *Proceedings of the National academy of Sciences of the United States of America* 108: 10092–10097.
213. Luo, M.L., et al. 2016. The CRISPR RNA-guided surveillance complex in Escherichia coli accommodates extended RNA spacers. *Nucleic Acids Research* 44: 7385–7394.
214. Craig, N. L., Craigie, R., Gellert, M. & Lambowitz, A. M. Mobile DNA III (2014).
215. Xu, H., S. Jia, and H. Xu. 2019. Therapeutic potential of exosomes in autoimmune diseases. *Clinical Immunology* 205: 116–124.
216. Rokad, D., H. Jin, V. Anantharam, A. Kanthasamy, and A.G. Kanthasamy. 2019. Exosomes as mediators of chemical-induced toxicity. *Curr Environ Health Rep.* 6 (3): 73–79.
217. Wang, Y., Z. Liu, X. Wang, Y. Dai, X. Li, S. Gao, et al. 2019. Rapid and quantitative analysis of exosomes by a chemiluminescence immunoassay using superparamagnetic iron oxide particles. *Journal of Biomedical Nanotechnology* 15 (8): 1792–1800.
218. Gao, M.L., F. He, B.C. Yin, and B.C. Ye. 2019. A dual signal amplification method for exosome detection based on DNA dendrimer self-assembly. *The Analyst* 144 (6): 1995–2002.
219. Houalla, R., F. Devaux, A. Fatica, J. Kufel, D. Barrass, C. Torchet, et al. 2006. Microarray detection of novel nuclear RNA substrates for the exosome. *Yeast* 23 (6): 439–454.
220. Wang, Q., L. Zou, X. Yang, X. Liu, W. Nie, Y. Zheng, et al. 2019. Direct quantification of cancerous exosomes via surface plasmon resonance with dual gold nanoparticle-assisted signal amplification. *Biosensors & Bioelectronics* 135: 129–136.

221. Rest, J.S., et al. 2013. Nonlinear fitness consequences of variation in expression level of a eukaryotic gene. *Molecular Biology and Evolution* 30: 448–456.
222. Bauer, C.R., S. Li, and M.L. Siegal. 2015. Essential gene disruptions reveal complex relationships between phenotypic robustness, pleiotropy, and fitness. *Molecular Systems Biology* 11: 773–773.
223. Keren, L., et al. 2016. Massively parallel interrogation of the effects of gene expression levels on fitness. *Cell* 166: 1282-1294.e18.
224. Dykhuizen, D.E., A.M. Dean, and D.L. Hartl. 1987. Metabolic flux and fitness. *Genetics* 115: 25–31.
225. Dekel, E., and U. Alon. 2005. Optimality and evolutionary tuning of the expression level of a protein. *Nature* 436: 588–592.
226. Alper, H., C. Fischer, E. Nevoigt, and G. Stephanopoulos. 2005. Tuning genetic control through promoter engineering. *Proceedings of the National academy of Sciences of the United States of America* 102: 12678–12683.
227. Perfeito, L., Ghozzi, S., Berg, J., Schnetz, K. & Lässig, M. Nonlinear fitness landscape of a molecular pathway. PLoS Genet. 7, e1002160 (2011).
228. Michaels, Y.S., et al. 2019. Precise tuning of gene expression levels in mammalian cells. *Nature Communications* 10: 818.
229. Patwardhan, R.P., et al. 2009. High-resolution analysis of DNA regulatory elements by synthetic saturation mutagenesis. *Nature Biotechnology* 27: 1173–1175.
230. Horlbeck, M. A. et al. Compact and highly active next-generation libraries for CRISPR-mediated gene repression and activation. eLife 5, e19760 (2016).
231. Mateescu, B., et al. 2017. Obstacles and opportunities in the functional analysis of extracellular vesicle RNA - An ISEV position paper. *J. Extracell. Vesicles.* https://doi.org/10.1080/200 13078.2017.1286095.
232. Zomer, A., et al. 2015. In vivo imaging reveals extracellular vesicle-mediated phenocopying of metastatic behavior. *Cell* 161: 1046–1057.
233. Hung, M.E., and J.N. Leonard. 2016. A platform for actively loading cargo RNA to elucidate limiting steps in EV-mediated delivery. *J. Extracell. Vesicles.* https://doi.org/10.3402/jev.v5. 31027.
234. Wang, H., et al. 2013. One-step generation of mice carrying mutations in multiple genes by CRISPR/Cas-mediated genome engineering. *Cell* 153: 910–918.
235. Wang, L., et al. 2015. Large genomic fragment deletion and functional gene cassette knock-in via Cas9 protein mediated genome editing in one-cell rodent embryos. *Science and Reports* 5: 17517.
236. Zhou, J., et al. 2014. Dual sgRNAs facilitate CRISPR/Cas9-mediated mouse genome targeting. *FEBS Journal* 281: 1717–1725.
237. Fujii, W., Kawasaki, K., Sugiura, K. & Naito, K. Efficient generation of large-scale genome-modified mice using gRNA and CAS9 endonuclease. Nucleic Acids Res. 41, e187 (2013).
238. Hara, S. et al. Microinjection-based generation of mutant mice with a double mutation and a 0.5 Mb deletion in their genome by the CRISPR/Cas9 system. J. Reprod. Dev. 62, 531–536 (2016).
239. Seruggia, D., A. Fernandez, M. Cantero, P. Pelczar, and L. Montoliu. 2015. Functional validation of mouse tyrosinase non-coding regulatory DNA elements by CRISPR-Cas9-mediated mutagenesis. *Nucleic Acids Research* 43: 4855–4867.
240. Yen, S.T., M. Zhang, J.M. Deng, S.J. Usman, C.N. Smith, J. Parker-Thornburg, P.G. Swinton, J.F. Martin, and R.R. Behringer. 2014. Somatic mosaicism and allele complexity induced by CRISPR/Cas9 RNA injections in mouse zygotes. *Developmental Biology* 393: 3–9.
241. Han, Y., O.J. Slivano, C.K. Christie, A.W. Cheng, and J.M. Miano. 2015. CRISPR-Cas9 genome editing of a single regulatory element nearly abolishes target gene expression in mice–brief report. *Arteriosclerosis, Thrombosis, and Vascular Biology* 35: 312–315.
242. Sung, Y.H., et al. 2014. Highly efficient gene knockout in mice and zebrafish with RNA-guided endonucleases. *Genome Research* 24: 125–131.

243. Zhang, L. et al. Large genomic fragment deletions and insertions in mouse using CRISPR/Cas9. PLoS ONE 10, e0120396 (2015).

244. Bolukbasi, M.F., A. Gupta, and S.A. Wolfe. 2016. Creating and evaluating accurate CRISPR-Cas9 scalpels for genomic surgery. *Nature Methods* 13: 41–50.

245. Hsu, P.D., et al. 2013. DNA targeting specificity of RNA-guided Cas9 nucleases. *Nature Biotechnology* 31: 827–832.

246. He, X. et al. Knock-in of large reporter genes in human cells via CRISPR/Cas9-induced homology-dependent and independent DNA repair. Nucleic Acids Res. 44, e85 (2016).

247. Lombardo, A., et al. 2007. Gene editing in human stem cells using zinc finger nucleases and integrase-defective lentiviral vector delivery. *Nature Biotechnology* 25: 1298–1306.

248. Lin, S., Staahl, B. T., Alla, R. K. & Doudna, J. A. Enhanced homology-directed human genome engineering by controlled timing of CRISPR/Cas9 delivery. Elife 3, e04766 (2014).

249. Zwaka, T.P., and J.A. Thomson. 2003. Homologous recombination in human embryonic stem cells. *Nature Biotechnology* 21: 319–321.

250. Hockemeyer, D., et al. 2009. Efficient targeting of expressed and silent genes in human ESCs and iPSCs using zinc-finger nucleases. *Nature Biotechnology* 27: 851–857.

251. Liu, Y., and M. Rao. 2011. Gene targeting in human pluripotent stem cells. *Methods in Molecular Biology* 767: 355–367.

252. Hockemeyer, D., and R. Jaenisch. 2016. Induced pluripotent stem cells meet genome editing. *Cell Stem Cell* 18: 573–586.

253. Song, H., S.K. Chung, and Y. Xu. 2010. Modeling disease in human ESCs using an efficient BAC-based homologous recombination system. *Cell Stem Cell* 6: 80–89.

254. Merkle, F.T., et al. 2015. Efficient CRISPR-Cas9-mediated generation of knockin human pluripotent stem cells lacking undesired mutations at the targeted locus. *Cell Reports* 11: 875–883.

255. Canny, M.D., et al. 2018. *Nature Biotechnology* 36: 95–102.

256. Cuella-Martin, R., et al. 2016. *Molecular Cell* 64: 51–64.

257. Doench, J.G., et al. 2016. *Nature Biotechnology* 34: 184–191.

258. Tsai, S.Q., et al. 2015. GUIDE-seq enables genome-wide profiling of off-target cleavage by CRISPR-Cas nucleases. *Nature Biotechnology* 33: 187–197.

259. Koike-Yusa, H., Y. Li, E.P. Tan, M.D.C. Velasco-Herrera, and K. Yusa. 2014. Genome-wide recessive genetic screening in mammalian cells with a lentiviral CRISPR-guide RNA library. *Nature Biotechnology* 32: 267–273.

260. van Overbeek, M., et al. 2016. DNA repair profiling reveals nonrandom outcomes at Cas9-mediated breaks. *Molecular Cell* 63: 633–646.

261. Tan, E.P., Y. Li, M.D.C. Velasco-Herrera, K. Yusa, and A. Bradley. 2015. Off-target assessment of CRISPR-Cas9 guiding RNAs in human iPS and mouse ES cells. *Genesis* 53: 225–236.

262. Piras, F., et al. 2017. Lentiviral vectors escape innate sensing but trigger p53 in human hematopoietic stem and progenitor cells. *EMBO Molecular Medicine* 9: 1198–1211.

263. Zacharias, J., L.G. Romanova, J. Menk, and N.J. Philpott. 2011. p53 inhibits adeno-associated viral vector integration. *Human Gene Therapy* 22: 1445–1451.

264. Haapaniemi, E., S. Botla, J. Persson, B. Schmierer, and J. Taipale. 2018. CRISPR–Cas9 genome editing induces a p53-mediated DNA damage response. *Nature Medicine* 24: 927–930.

265. Ihry, R.J., et al. 2018. p53 inhibits CRISPR–Cas9 engineering in human pluripotent stem cells. *Nature Medicine* 24: 939–946.

266. Schiroli, G., et al. 2019. Precise gene editing preserves hematopoietic stem cell function following transient p53-mediated DNA damage response. *Cell Stem Cell* 24: 551-565.e8.

267. Giacomelli, A.O., et al. 2018. Mutational processes shape the landscape of TP53 mutations in human cancer. *Nature Genetics* 50: 1381–1387.

Chapter 3
Extension and Improvement of CRISPR-Based Technology

Ziheng Zhang, Ping Wang, and Ji-Long Liu

Introduction

The initial use of CRISPR/Cas9 was based on its ability to cut the target DNA sequence. The advent of CRISPR provides researchers with the possibility of editing almost any DNA sequence. CRISPR, however, gives researchers a general paradigm that targets specific DNA sequences. After breaking through the fixed mindset that 'CRISPR is a tool for cutting DNA', a series of CRISPR-based toolkits were developed. The use of CRISPR-related technologies is no longer limited to knocking out target genes.

Because CRISPR can accurately bind to the target DNA or RNA sequences, a series of edits can be made on the locus of interest; for example, CRISPR-based inhibition or activation of gene expression by fusing inhibiter peptides or activator peptides on dCas9; CRISPR-based epigenetic modification by fusing an epigenetic modifier on dCas9; CRISPR-based DNA or RNA tracking and imaging by fusing fluorescent protein on dCas9 or dCas13; CRISPR-based DNA/RNA–protein interactions capturing by fusing proximity label enzymes; and CRISRP-based base editing by fusing a cytidine deaminase or adenine deaminase. In this section, we mainly focus on a series of CRISPR-based toolkits for functional expansion. These expansions include the application of other members of the Cas family, the engineering of Cas nucleases and the use of CRISPR for other aspects besides gene knockout. In particular, we regard the base editor as a separate section owing to its unique working method and huge application prospects.

Z. Zhang
School of Life Science and Technology, ShanghaiTech University, Shanghai, China

P. Wang
University Library, ShanghaiTech University, Shanghai, China

J.-L. Liu (✉)
School of Life Science and Technology, ShanghaiTech University, Shanghai, China

Machine generated keywords: variant, spcas, pam, base editor, editor, offtarget, grna, cytosine, deaminase, deamination, predict, base, apobec, sgrna, casa.

3.1   Engineering and Application of Cas Nuclease

Machine generated keywords: spcas, variant, pam, pam sequence, expand, casa, cut, amplification, requirement, evolve, phage, epigenetic, offtarget, grna, dca.

Epigenome editing by a CRISPR-Cas9-based acetyltransferase activates genes from promoters and enhancers.

https://doi.org/10.1038/nbt.3199

Main

Technologies for targeted direct manipulation of epigenetic marks are needed to transform association-based findings into mechanistic principles of gene regulation.

Fusions of epigenome-modifying enzymes to programmable DNA-binding proteins, such as zinc finger proteins and transcription activator-like effectors (TALEs), are effective at achieving targeted DNA methylation and hydroxymethylation, and histone demethylation, methylation and deacetylation [1–6].

A method for targeted histone acetylation, which is strongly associated with active gene regulatory elements and enhancers, has not been described.

The CRISPR-Cas9 (clustered, regularly interspaced, short palindromic repeat–CRISPR-associated protein) genome engineering tool [7, 8], which can be readily targeted to loci of interest, has not yet been extensively applied to epigenome editing.

The Cas9 nuclease can be directed to specific genomic loci using complementarity between an engineered guide RNA (gRNA) and the target site [9–11].

We hypothesized that recruitment of an acetyltransferase by dCas9 to a genomic target site would directly modulate the epigenome and activate nearby gene expression.

Results

Although there are many published examples of genes being activated with engineered transcription factors targeted to promoters, inducing gene expression from other distal regulatory elements has been limited, particularly for dCas9-based activators [12–15].

Given the role and localization of p300 at endogenous enhancers [16, 17], we hypothesized that the dCas9$^{p300\text{ Core}}$ would effectively induce transcription from distal regulatory regions with appropriately targeted gRNAs.

DCas9$^{p300\text{ Core}}$ induced significant transcription when targeted to either MYOD regulatory element with corresponding gRNAs (P = 0.0115 and 0.0009 for the CE and DRR regions, respectively).

Targeting of dCas9$^{p300\ Core}$ to the HS2 enhancer led to significant expression of the downstream HBE, HBG and HBD genes ($P \leq 0.0001$, 0.0056, and 0.0003 between dCas9$^{p300\ Core}$ and mock-transfected cells for HBE, HBG and HBD, respectively).

With the exception of the most distal HBB gene, dCas9$^{p300\ Core}$ activated transcription from downstream genes when targeted to all characterized enhancer regions assayed, a capability not observed for dCas9^{VP64}.

Discussion

These results establish the dCas9$^{p300\ Core}$ fusion protein as a potent and easily programmable tool to synthetically manipulate acetylation at targeted endogenous loci, leading to regulation of proximal and distal enhancer–regulated genes.

The observation that targeted acetylation is sufficient for gene activation at an endogenous locus and enhancer is a notable finding of our study.

The unique activity of dCas9$^{p300\ Core}$ supports its use in elucidating key steps of gene regulation, including dissection of the interplay between the epigenome, regulatory element activity and gene regulation.

The observed potent activation of globin gene expression by dCas9$^{p300\ Core}$-mediated acetylation in the absence of GATA1 suggests that the acetylation at the enhancer occurs downstream of GATA1 but upstream of globin gene activation in erythroid cells.

dCas9$^{p300\ Core}$ takes advantage of the simple programmability of the CRISPR-Cas9 system to target acetyltransferase activity and complements other recently described epigenetic editing tools, including fusions of demethylases, methyltransferases and deacetylases [1–5] to generate a more complete set of epigenome editing tools.

Methods

Transfections were performed in 24-well plates using 375 ng of respective dCas9 expression vector and 125 ng of equimolar pooled or individual gRNA expression vectors mixed with Lipofectamine 2000 (Life Technologies, cat #11668019) as per manufacturer's instruction.

For ChIP-qPCR experiments, HEK293T cells were transfected in 15-cm dishes with Lipofectamine 2000 and 30 μg of respective dCas9 expression vector and 10 μg of equimolar pooled gRNA expression vectors as per manufacturer's instruction.

An HA epitope tag was added to dCas9 (no effector) by removing the VP64 effector domain from dCas9^{VP64} via AscI/PacI restriction sites and using isothermal assembly [18] to include an annealed set of oligos containing the appropriate sequence as per manufacturer's instruction (NEB cat #2611).

An AscI site, HA-epitope tag, and a PmeI site were added by PCR amplification of the p300 Core from pCR-Blunt$^{p300\ Core}$, and subsequently this amplicon was cloned into pCR-Blunt (pCR-Blunt$^{p300\ Core\ +\ HA}$) (Life Technologies cat #K2700).

Acknowledgements

A machine generated summary based on the work of Hilton, Isaac B; D'Ippolito, Anthony M; Vockley, Christopher M; Thakore, Pratiksha I; Crawford, Gregory E; Reddy, Timothy E; Gersbach, Charles A 2015 in Nature Biotechnology.

High-fidelity CRISPR–Cas9 nucleases with no detectable genome-wide off-target effects.

https://doi.org/10.1038/nature16526

Abstract-Summary

Existing strategies for reducing genome-wide off-target effects of the widely used Streptococcus pyogenes Cas9 (SpCas9) are imperfect, possessing only partial or unproven efficacies and other limitations that constrain their use.

SpCas9-HF1 retains on-target activities comparable to wild-type SpCas9 with >85% of single-guide RNAs (sgRNAs) tested in human cells.

With sgRNAs targeted to standard non-repetitive sequences, SpCas9-HF1 rendered all or nearly all off-target events undetectable by genome-wide break capture and targeted sequencing methods.

Even for atypical, repetitive target sites, the vast majority of off-target mutations induced by wild-type SpCas9 were not detected with SpCas9-HF1.

Main

Various strategies have been described to reduce genome-wide off-target mutations of the commonly used SpCas9 nuclease, including: truncated sgRNAs bearing shortened regions of target site complementarity [19, 20], SpCas9 mutants such as the recently described D1135E variant [21], paired SpCas9 nickases [22, 23], and dimeric fusions of catalytically inactive SpCas9 to a non-specific FokI nuclease [24–26].

These approaches are only partially effective, have as-yet unproven efficacies on a genome-wide scale, and/or possess the potential to create more new off-target sites.

SpCas9–sgRNA complexes cleave target sites composed of an NGG protospacer adjacent motif (PAM) sequence (recognized by SpCas9) [10, 27–29] and an adjacent 20 base pair (bp) protospacer sequence (which is complementary to the 5′ end of the sgRNA) [9–11, 30].

We previously proposed that the SpCas9–sgRNA complex might possess more energy than is needed for optimal recognition of its intended target DNA site, thereby enabling cleavage of mismatched off-target sites [20].

Alteration of SpCas9 DNA contacts

Alanine substitution of one or all of these residues did not reduce on-target cleavage efficiency of SpCas9 with this EGFP-targeted sgRNA.

SpCas9-HF1 retains high on-target activities

SpCas9-HF1 possesses comparable activities (greater than 70% of wild-type SpCas9 activities) for 86% (32/37) of the sgRNAs we tested.

Genome-wide specificity of SpCas9-HF1

To test whether SpCas9-HF1 exhibits reduced off-target effects in human cells, we used the genome-wide unbiased identification of double-stranded breaks enabled by sequencing (GUIDE-seq) method [19] to assess eight different sgRNAs targeted to sites in the endogenous human EMX1, FANCF, RUNX1, and ZSCAN2 genes.

To confirm these GUIDE-seq findings, we used targeted amplicon sequencing to more directly measure the frequencies of indel mutations induced by wild-type SpCas9 and SpCas9-HF1.

We used next-generation sequencing to examine the on-target sites and 36 of the 40 off-target sites that had been identified for six sgRNAs with wild-type SpCas9 in our GUIDE-seq experiments (four of the 40 sites could not be specifically amplified from genomic DNA).

We conclude that SpCas9-HF1 can completely or nearly completely reduce off-target mutations that occur across a range of different frequencies with wild-type SpCas9 to levels generally undetectable by GUIDE-seq and targeted deep sequencing.

Refining the specificity of SpCas9-HF1

We next sought to determine whether SpCas9-HF2, -HF3, or -HF4 could reduce indel frequencies at two off-target sites that remained susceptible to modification by SpCas9-HF1, one with the FANCF site 2 sgRNA and another with the VEGFA site 3 sgRNA.

Discussion

The SpCas9-HF1 variant characterized in this report reduces all or nearly all genome-wide off-target effects to undetectable levels as judged by GUIDE-seq and targeted next-generation sequencing, with the most robust and consistent effects observed with sgRNAs designed against standard, non-repetitive target sequences.

We found that introducing substitutions at other non-specific DNA contacting residues can further reduce some of the very small number of residual off-target sites that persist for certain sgRNAs with SpCas9-HF1.

Our variants might be combined with substitutions in residues that contact the non-target DNA strand, alterations that have been shown to reduce SpCas9 off-target effects while our manuscript was under review [31].

Methods

For T7 endonuclease I assays, GUIDE-seq experiments, and targeted deep sequencing, genomic DNA was extracted ~72 h post-transfection using the Agencourt DNAdvance Genomic DNA Isolation Kit (Beckman Coulter Genomics).

EGFP disruption experiments, in which cleavage and induction of indels by non-homologous end-joining (NHEJ)-mediated repair within a single integrated EGFP reporter gene leads to loss of cell fluorescence, were performed as previously described [32, 33].

GUIDE-seq relies on the integration of a short dsODN tag into DNA breaks to enable amplification and sequencing of adjacent genomic sequence, with the number of tag integrations at any given site providing a quantitative measure of cleavage efficiency [19].

U2OS cells were transfected with 750 ng of Cas9 and 250 ng sgRNA plasmids as described above, along with 100 pmol of a GUIDE-seq end-protected dsODN that contains an NdeI restriction site [19].

PCR products were generated for each on- and off-target site from ~100 ng of genomic DNA extracted from U2OS cells.

Acknowledgements

A machine generated summary based on the work of Kleinstiver, Benjamin P.; Pattanayak, Vikram; Prew, Michelle S.; Tsai, Shengdar Q.; Nguyen, Nhu T.; Zheng, Zongli; Joung, J. Keith 2016 in Nature.

Enhancing homology-directed genome editing by catalytically active and inactive CRISPR-Cas9 using asymmetric donor DNA.

https://doi.org/10.1038/nbt.3481

Main

After binding, the HNH and RuvC nuclease domains within Cas9 respectively cleave the target and nontarget strands of substrate DNA.

Direct comparisons between our in vitro data and the timing of DNA repair in cells could be complicated by many factors, but cellular single-molecule measurements of Cas9 binding also indicate an unusually long residence time at on-target sites [34], consistent with a model in which Cas9 stably binds to substrate DNA, concealing the underlying double-strand break and preventing recognition by genome surveillance factors.

To results from previous studies, which focused on donor DNA symmetric around the break [35–37], we observed that asymmetric donor DNA optimized for annealing by overlapping the Cas9 cut site with 36 bp on the PAM-distal side, and with a 91-bp extension on the PAM-proximal side of the break, supported HDR frequencies of 57 ± 5%.

Methods

200 nM Cas9 (expressed from pCR1002) or dCas9 (expressed from pCR1003) was mixed with a 20% molar excess of sgRNA and incubated for 10 min to form RNP.

Biosensor-dsDNA-RNP complexes were allowed to dissociate in Reaction Buffer for 3600 s. Response curves for each biosensor were normalized against biosensors conjugated to DNA but without RNP (buffer-only control).

Nuclease and sgRNA were incubated in Reaction Buffer for 30 min to form RNP.

200 ng of genomic DNA was used as a template in a PCR reaction using KAPA high GC buffer (KAPA Biosystems, Wilmington, MA) and standard PCR conditions

(95 °C for 5 min, 35 cycles of 98 °C for 20 s, 62 °C for 15 s and 72 °C for 30 s, and one cycle of 72 °C for 1 min).

Acknowledgements

A machine generated summary based on the work of Richardson, Christopher D; Ray, Graham J; DeWitt, Mark A; Curie, Gemma L; Corn, Jacob E 2016 in Nature Biotechnology.

In vivo *genome editing with a small Cas9 orthologue derived from Campylobacter jejuni.*

https://doi.org/10.1038/ncomms14500

Abstract-Summary

We present the smallest Cas9 orthologue characterized to date, derived from Campylobacter jejuni (CjCas9), for efficient genome editing in vivo.

CjCas9, delivered via AAV, induces targeted mutations at high frequencies in mouse muscle cells or retinal pigment epithelium (RPE) cells.

CjCas9 targeted to the Vegfa or Hif1a gene in RPE cells reduces the size of laser-induced choroidal neovascularization, suggesting that in vivo genome editing with CjCas9 is a new option for the treatment of age-related macular degeneration.

Extended:

We present Campylobacter jejuni-derived Cas9 (CjCas9) for efficient genome editing in vitro and in vivo.

The data that support the findings of this study are available from the corresponding author upon reasonable request.

Introduction

Cas9 RGENs cleave chromosomal DNA in a targeted manner, enabling genetic modifications or genome editing in cells and whole organisms [9, 11, 38–41].

Cas9 derived from Streptococcus pyogenes (SpCas9), the first Cas9 orthologue to enable targeted mutagenesis in human cells [9, 11, 30, 38], is still the most widely used among several Cas9 proteins available for genome editing.

The SpCas9 gene alone can be packaged into a single AAV vector in vivo [42, 43].

To SpCas9, Staphylococcus aureus Cas9 (SaCas9) can be used for genome editing [44].

Smaller Cas9 orthologues with different PAM sequences are highly desired to expand in vivo genome editing.

We present Campylobacter jejuni-derived Cas9 (CjCas9) for efficient genome editing in vitro and in vivo.

Results

These reporter assays showed that CjCas9 cleaved target sites containing 5′-NNNNRYAC-3′ PAM sequences in HEK 293 cells.

One target site in the human genome contained a 5′-NGGNACAC-3′ PAM recognized by both CjCas9 and SpCas9.

Two CjCas9 nucleases targeting the Rosa26 locus or the Vegfa gene in the mouse genome cleaved genomic DNA only at the single on-target site, reminiscent of the remarkable specificity of Cpf1 nucleases [45, 46].

We chose 3 overlapping sites in the human genome with 5′-NNGRRTAC-3′ PAM sequences, which can be targeted by both CjCas9 and SaCas9.

CjCas9 and SaCas9 targeted to the AAVS1-TS16 site cleaved human genomic DNA at 59 and 118 sites, respectively.

CjCas9 and SaCas9 targeted to the AAVS1-TS39 site were more specific than were those targeted to the TS16 site, cleaving human genomic DNA at 2 and 15 sites, respectively.

Discussion

We showed that CjCas9 targeted to the Hif1a gene in mouse eyes inactivated the gene in RPE cells efficiently and reduced the area of CNV in a mouse model of AMD.

Because the CjCas9 target site in the mouse Hif1a gene is perfectly conserved in the human HIF1A gene, the AAV presented in this study or its variants could be used for the treatment of human patients in the future.

We expect that CjCas9 can be directed to other traditionally 'undruggable' genes or non-coding sequences to broaden the range of therapeutic targets, making the entire human genome potentially druggable.

Methods

To determine optimal PAM sequences, each of the resulting reporter plasmids (100 ng) and plasmids encoding CjCas9 and its sgRNA (225 and 675 ng, respectively) were co-transfected into HEK293 cells (1×10^5) using lipofectamine 2000 (Invitrogen).

Genomic DNA (8 μg) with CjCas9 or SaCas9 protein (300 nM) and sgRNA (900 nM) in a 400 μl reaction volume (100 mM NaCl, 50 mM Tris–HCl, 10 mM MgCl$_2$, and 100 μg ml^{-1} BSA) and incubated the mixture for 8 h at 37 °C.

AAV inverted terminal repeat-based vector plasmids carrying a sgRNA sequence and the CjCas9 gene with a nuclear localization signal and an HA tag at the C terminus were constructed.

The opsin-positive area corresponding to RPE cells expressing HA-tagged CjCas9 was measured using Image J software (1.47v, NIH) by blinded observers.

The CjCas9 protein expressed in TA muscles of C57BL/6 J mice at 8 months after injection of AAV was detected using western blotting.

Additional information

How to cite this article: Kim, E. and others In vivo genome editing with a small Cas9 orthologue derived from Campylobacter jejuni.

Publisher's note: Springer Nature remains neutral with regard to jurisdictional claims in published maps and institutional affiliations.

Acknowledgements

A machine generated summary based on the work of Kim, Eunji; Koo, Taeyoung; Park, Sung Wook; Kim, Daesik; Kim, Kyoungmi; Cho, Hee-Yeon; Song, Dong Woo; Lee, Kyu Jun; Jung, Min Hee; Kim, Seokjoong; Kim, Jin Hyoung; Kim, Jeong Hun; Kim, Jin-Soo 2017 in Nature Communications.

Evolved Cas9 variants with broad PAM compatibility and high DNA specificity.

https://doi.org/10.1038/nature26155

Abstract-Summary

For the most commonly used Cas9 from Streptococcus pyogenes (SpCas9), the required PAM sequence is NGG.

The PAM compatibility of xCas9 is the broadest reported, to our knowledge, among Cas9 proteins that are active in mammalian cells, and supports applications in human cells including targeted transcriptional activation, nuclease-mediated gene disruption, and cytidine and adenine base editing.

Despite its broadened PAM compatibility, xCas9 has much greater DNA specificity than SpCas9, with substantially lower genome-wide off-target activity at all NGG target sites tested, as well as minimal off-target activity when targeting genomic sites with non-NGG PAMs.

These findings expand the DNA targeting scope of CRISPR systems and establish that there is no necessary trade-off between Cas9 editing efficiency, PAM compatibility and DNA specificity.

Main

To a protospacer that complements the sgRNA, a Cas9 target site must also contain a PAM sequence to support recognition by Cas9.

The NGG PAM requirement of canonical SpCas9, which occurs on average only once in every 16 randomly chosen genomic loci, greatly limits the targeting scope of Cas9 especially for applications that require precise Cas9 positioning, such as base editing, which requires a PAM approximately 13–17 nucleotides from the target base [47, 48], and some forms of homology-directed repair, which are most efficient when DNA cleavage occurs roughly 10–20 base pairs away from a desired alteration [36, 49].

To address this limitation, researchers have harnessed natural CRISPR nucleases with different PAM requirements and engineered existing systems to accept variants of naturally recognized PAMs.

Although CRISPR nucleases engineered to accept additional PAM sequences [21, 50] also expand the scope of genomic targets available for Cas9-mediated manipulation, many target sequences remain inaccessible.

We used phage-assisted continuous evolution (PACE) to rapidly generate Cas9 variants that accept an expanded range of PAM sequences.

Evolution of Cas9 towards expanded PAM compatibility

We envisioned installing a library of all 64 possible NNN PAM sequences at the target protospacer in the accessory plasmid, so that selection phages encoding Cas9 variants with broader PAM compatibility would replicate in a larger fraction of host cells and thus experience a fitness advantage.

The mixture of phage from the final phage-assisted non-continuous evolution pool were further evolved for 72 h using PACE on host cells containing the same accessory plasmid libraries harbouring NNN PAM sequences.

The resulting phages were continuously evolved using PACE for an additional 72 h on host cells containing three protospacer–sgRNA pairs and HHH PAM libraries, in which H is A, C, or T, to favour Cas9 variants with activity on non-NGG PAMs.

The xCas9-3.0–xCas9-3.13 clones were tested in a PAM depletion assay [21, 50] in which they were given the opportunity to cleave a library of plasmids containing a protospacer and all possible NNN PAM sequences in an antibiotic-resistance gene in bacterial cells.

xCas9 activators and nucleases in human cells

To test targeted genomic DNA cleavage in human cells, we expressed xCas9-3.7 and -3.6 nuclease in a HEK293T cell line with a genomically integrated GFP gene and measured the loss of GFP fluorescence reflecting DNA cleavage and indel-mediated disruption of the target site.

Neither xCas9-3.7 nor xCas9-3.6 increased GFP loss relative to SpCas9 for either NNG PAM site tested, suggesting that the transcriptional activation by dxCas9–VPR observed at some NNG PAM sites, and the strong NNG PAM signal in the bacterial PAM depletion assay, do not necessarily translate to DNA cleavage at all NNG sites in mammalian cells.

Among the three GAA and GAT sites tested, SpCas9 showed virtually no activity, averaging $1.4 \pm 1.3\%$ indel formation, whereas xCas9-3.7 averaged $7.2 \pm 2.8\%$ indel formation, a 5.2-fold increase.

Base editing by xCas9 variants in human cells

These results indicate that xCas9 variants are compatible with the BE3 architecture enabling cytidine base editing of target sites that cannot be accessed by SpCas9–BE3 or, with the exception of NGA PAM sites [51], by any other previously reported base editors.

Average base-editing efficiency at the NGG PAM site tested increased from $48 \pm 2.1\%$ to $69 \pm 3.7\%$.

At the GAT PAM site tested, xCas9(3.7)–ABE resulted in $16 \pm 1.5\%$ base editing, whereas SpCas9–ABE yielded no detectable editing (at most 0.1%), representing a more than 100-fold increase.

On the two NGC and three NGA sites tested, xCas9(3.7)–ABE averaged $21 \pm 2.5\%$ and $43 \pm 1.5\%$ base editing, respectively, whereas SpCas9–ABE averaged $7.0 \pm 1.3\%$ and $22 \pm 1.2\%$, respectively.

Improved DNA specificity of xCas9 in human cells

We also evaluated the off-target DNA specificity of both SpCas9 and xCas9-3.7 at two non-NGG PAM (GAA and CGT) sites in HEK293T cells.

That xCas9 exhibits much higher DNA specificity than SpCas9 even though it was not explicitly selected for this property suggests that the off-target activity of wild-type SpCas9 may lie at a narrow fitness peak that is suitable for defending the much smaller bacterial genome but not optimal for genome editing in mammalian cells.

Evolved xCas9 variants are also the first to offer improvements in targeting scope, editing efficiency and DNA specificity in a single entity relative to wild-type SpCas9.

Although the efficacy of xCas9 on non-NGG PAMs varies based on application (transcriptional activation, DNA cleavage or base editing) and on target site, the ability to access some NG, GAA and GAT PAM sequences greatly expands the breadth of targets available for site-sensitive genome editing applications.

Methods

PCR was performed using Q5 Hot Start High-Fidelity Polymerase (New England Biolabs) with phosphorylated primers and the plasmid pFYF1320 (sgRNA expression plasmid with a spacer for targeting enhanced GFP (eGFP)) as a template according to the manufacturer's instructions.

For genomic DNA cutting or base editing, 750 ng of Cas9 or BE3 and 250 ng of sgRNA expression plasmids were transfected using 1.5 μl of Lipofectamine 2000 (Thermo Fisher Scientific) per well according to the manufacturer's protocol.

For GFP activation, 200 ng of dCas9–VPR plasmid, 50 ng of sgRNA expression plasmid, 60 ng of GFP reporter plasmid and 30 ng of near-infrared fluorescent protein (iRFP) expression plasmid were transfected using 1.5 μl Lipofectamine 2000 (Thermo Fisher Scientific) per well according to the manufacturer's protocol.

Each 25 μl PCR reaction was assembled with Phusion Hot Start II High-Fidelity DNA Polymerase (Thermo Fisher Scientific) according to the manufacturer's instructions using 1.0 μM forward and reverse primer and 1 μl of genomic DNA extract

Acknowledgements

A machine generated summary based on the work of Hu, Johnny H.; Miller, Shannon M.; Geurts, Maarten H.; Tang, Weixin; Chen, Liwei; Sun, Ning; Zeina, Christina M.; Gao, Xue; Rees, Holly A.; Lin, Zhi; Liu, David R. 2018 in Nature.

Engineered CRISPR–Cas12a variants with increased activities and improved targeting ranges for gene, epigenetic and base editing

https://doi.org/10.1038/s41587-018-0011-0

Abstract-Summary

Cas12a variant (enAsCas12a) that has a substantially expanded targeting range, enabling targeting of many previously inaccessible PAMs.

EnAsCas12a exhibits a twofold higher genome editing activity on sites with canonical TTTV PAMs compared to wild-type AsCas12a, and we successfully grafted a subset of mutations from enAsCas12a onto other previously described AsCas12a variants [52] to enhance their activities.

enAsCas12a improves the efficiency of multiplex gene editing, endogenous gene activation and C-to-T base editing, and we engineered a high-fidelity version of enAsCas12a (enAsCas12a-HF1) to reduce off-target effects.

EnAsCas12a provides an optimized version of Cas12a that should enable wider application of Cas12a enzymes for gene and epigenetic editing.

Main

We examined the editing activities of the E174R/S542R/K548R variant on 97 other sites in human cells bearing 28 additional PAMs identified as targetable by the PAMDA.

To determine whether the increased activity phenotype of enAsCas12a might be transferable to other AsCas12a variants, we added the E174R substitution to the previously described RVR and RR PAM recognition variants [52] to create enRVR and enRR, respectively.

The enhanced AsCas12a variants described herein substantially improve the targeting range, on-target activities and fidelity of Cas12a nucleases, which are properties that are important for multiplex gene editing, epigenetic editing, cytosine base editing and gene knockout in primary human T cells.

enAsCas12a also exhibits superior on-target activity relative to wild-type AsCas12a, increasing editing efficiencies by approximately twofold on sites with canonical TTTV PAMs in two cell lines and in primary human T cells.

Methods

Cultures were then induced with 0.2 mM isopropyl β-D-thiogalactopyranoside before shaking at 18 °C for 23 h. Cell pellets from 50 ml of culture were harvested by centrifugation at 1200 g for 15 min and suspended in 1 ml lysis buffer v1 containing 20 mM Hepes pH 7.5, 100 mM KCl, 5 mM MgCl2, 5% glycerol, 1 mM DTT, Sigmafast protease inhibitor (Sigma-Aldrich) and 0.1% Triton X-100.

The cell lysate was centrifuged for 20 min at 21,000 g and 4 °C, and the supernatant was mixed with an equal volume of binding buffer v1 (lysis buffer v1 with 10 mM imidazole), added to 400 μl of HisPur Ni–NTA Resin (Thermo Fisher Scientific) that was pre-equilibrated in binding buffer v1 and rocked at 4 °C for 8 h. The protein-bound resin was washed three times with 1 ml wash buffer v1 (20 mM Hepes pH 7.5, 500 mM KCl, 5 mM $MgCl_2$, 5% glycerol, 25 mM imidazole and 0.1% Triton X-100) and then once with 1 ml binding buffer v1.

Acknowledgements

A machine generated summary based on the work of Kleinstiver, Benjamin P.; Sousa, Alexander A.; Walton, Russell T.; Tak, Y. Esther; Hsu, Jonathan Y.; Clement, Kendell; Welch, Moira M.; Horng, Joy E.; Malagon-Lopez, Jose; Scarfò, Irene; Maus, Marcela V.; Pinello, Luca; Aryee, Martin J.; Joung, J. Keith 2019 in Nature Biotechnology.

Detection of unamplified target genes via CRISPR–Cas9 immobilized on a graphene field-effect transistor.

https://doi.org/10.1038/s41551-019-0371-x

Abstract-Summary

Termed CRISPR–Chip, the biosensor uses the gene-targeting capacity of catalytically deactivated CRISPR-associated protein 9 (Cas9) complexed with a specific single-guide RNA and immobilized on the transistor to yield a label-free nucleic-acid-testing device whose output signal can be measured with a simple handheld reader.

In the presence of genomic DNA containing the target gene, CRISPR–Chip generates, within 15 min, with a sensitivity of 1.7 fM and without the need for amplification, a significant enhancement in output signal relative to samples lacking the target sequence.

CRISPR–Chip expands the applications of CRISPR–Cas9 technology to the on-chip electrical detection of nucleic acids.

Extended:

Future work invites more studies to enhance the capability of CRISPR–Chip to detect single nucleotide polymorphisms, which could significantly broaden its future clinical application.

Main

This expanded knowledge has allowed for the development of various target-specific nucleic acid detection tools [53–55].

Clustered regularly interspaced short palindromic repeats (CRISPR)-associated nuclease (Cas)-based methodologies have been utilized to improve on conventional nucleic acid targeting for optical detection [56–58].

In 2017, Gootenberg and others [56] reported the development of the SHERLOCK methodology, which utilized an RNA-guided RNA targeting Cas13a (CRISPR-associated nuclease 13a) to provide a fluorescent signal readout after Cas13a complex hybridization with its target sequence, once initially amplified by recombinase polymerase amplification (RPA).

Both SHERLOCK and HOLMES allowed for the sequence-specific detection of DNA and RNA via collateral cleavage of a single-stranded nucleic acid probe post-amplification [59].

The graphene is functionalized with a catalytically deactivated Cas9 (dCas9) CRISPR complex, denoted as dRNP, which interacts with its target sequence by scanning the whole genomic sample, unzipping the double helix and associating upstream of a protospacer adjacent motif [60] until it finds and binds to the target sequence that is complementary to the single-guide RNA molecule (sgRNA) within the dRNP [29, 35].

Results

We investigated CRISPR–Chip's ability to detect whole-genome samples containing the bfp target.

To investigate CRISPR–Chip ability to detect the bfp gene within whole-genome samples, dRNP-BFP-functionalized CRISPR–Chips were fabricated and evaluated in the presence of varying concentrations of the target HEK-BFP genomic samples (300–1200 ng).

For control experiments, dRNP-BFP-functionalized CRISPR–Chips were incubated with 900 ng of HEK genomic sample, which lacked the bfp gene.

These results indicate a limit of detection (LOD) of 2.3 fM. In addition, the real-time monitoring of the I response indicates that after CRISPR–Chip incubation with genomic samples, the signal reached its saturation within 5 min, showing that this incubation time was sufficient, before rinsing the sensor to remove unbound genomic material, to obtain the final I response.

CRISPR–Chip's ability to detect DMD-associated mutations was evaluated using two dRNPs, which targeted exons 3 and 51 of the human dystrophin gene.

Discussion

The RNA-guided dCas9 complex immobilized on the surface of the graphene within in the CRISPR–Chip gFET construct can specifically bind, enrich and detect the target DNA without the need for reagents and bulky instruments.

CRISPR–Chip was able to specifically detect the deletion of two target sequences in DMD patients without any pre-amplification.

Although CRISPR–Chips were designed to detect only two common DMD exon deletions, these studies indicate that the targeting capacity of CRISPR–Chip can be expanded for multiplex gene analysis by simply modifying the unique 20-nucleotide sequence at the 5′ end of the sgRNA molecule within the dRNP construct.

The CRISPR–Chip LOD was lower than for previously reported amplification-free technologies for the detection of target sequences contained within the whole genome [55, 61, 62].

Other amplification methodologies such as RPA, used in other CRISPR-based diagnostics technologies, could be utilized to increase the target sequence copy number before CRISPR–Chip analysis [56, 63].

Methods

The dRNP-BFP-functionalized CRISPR–Chips were calibrated with 2 mM $MgCl_2$ for 5 min at 37 °C and subsequently incubated with varying concentrations (300–1200 ng) of HEK-BFP (30 µl in 2 mM $MgCl_2$) for 30 min at 37 °C.

The dRNP-BFP-functionalized CRISPR–Chips were calibrated with 2 mM $MgCl_2$ for 5 min at 37 °C and subsequently incubated with 900 ng HEK-BFP DNA mixed with varied concentrations (0–1800 ng) of HEK DNA (30 µl in 2 mM MgCl2).

The dRNP-DMD3- and dRNP-DMD51-functionalized CRISPR–Chips were calibrated with 2 mM $MgCl_2$ for 5 min at 37 °C and subsequently incubated with 900 ng (30 µl, 2 mM $MgCl_2$) genomic clinical sample.

For sensitivity evaluation, dRNP-DMD51-functionalized CRISPR–Chips were calibrated with 2 mM $MgCl_2$ for 5 min at 37 °C and subsequently incubated with varied concentrations of sample A (400–1500 ng) (30 µl, 2 mM $MgCl_2$).

Acknowledgements

A machine generated summary based on the work of Hajian, Reza; Balderston, Sarah; Tran, Thanhtra; deBoer, Tara; Etienne, Jessy; Sandhu, Mandeep; Wauford, Noreen A.; Chung, Jing-Yi; Nokes, Jolie; Athaiya, Mitre; Paredes, Jacobo; Peytavi, Regis; Goldsmith, Brett; Murthy, Niren; Conboy, Irina M.; Aran, Kiana 2019 in Nature Biomedical Engineering.

Targeted nanopore sequencing with Cas9-guided adapter ligation.

https://doi.org/10.1038/s41587-020-0407-5

Abstract-Summary

Despite recent improvements in sequencing methods, there remains a need for assays that provide high sequencing depth and comprehensive variant detection.

We describe nanopore Cas9-targeted sequencing (nCATS), an enrichment strategy that uses targeted cleavage of chromosomal DNA with Cas9 to ligate adapters for nanopore sequencing.

The nCATS sequencing requires only ~3 µg of genomic DNA and can target a large number of loci in a single reaction.

The method will facilitate the use of long-read sequencing in research and in the clinic.

Main

Targeted sequencing allows investigators to enrich for loci of interest, reducing sequencing costs and labor to achieve high coverage data at desired genomic regions.

In the GM12878 cell line with single gRNAs flanking each site, after quality-filtering alignments (MAPping Quality (MAPQ) > 30), there were only two genomic sites outside target regions where coverage reached 25 × . Both of these are at repetitive pericentromeric sites and contain reads with lower mapping quality (MAPQ = 30–50), suggesting the increased coverage to be the result of alignment errors in these poorly mappable regions.

Sites for methylation studies were selected by searching whole-genome nanopore data [64] for differentially methylated promoters between the non-tumorigenic breast cell line MCF-10A and the tumorigenic breast cell lines MCF-7 and MDA-MB-231.

We found that, by using only high-confidence variants, we could phase nanopore sequencing reads into parental alleles using WhatsHap [65], permitting haplotype resolution of high-coverage nanopore data.

Methods

B7204) at a final volume of 30 μl (concentration 333 nM), incubated for 20 min at room temperature, then stored at 4 °C until use, up to 2 d. Input DNA, 3 μg, was resuspended in 30 μl of 1 × CutSmart buffer (New England Biolabs, catalog no.

LSK109) before eluting in 15 μl of elution buffer (Oxford Nanopore Technologies, catalog no.

Sequencing libraries were prepared by adding the following to the eluate: 25 μl of sequencing buffer (Oxford Nanopore Technologies, catalog no.

LSK109) and 0.5 μl of sequencing tether (Oxford Nanopore Technologies, catalog no.

Base calling was performed using GUPPY (v.3.0.3) to generate FASTQ sequencing reads from the electrical data.

New variant calling was performed using Samtools (v.1.9) [66], Clair (v.2.0.0) [67], Medaka (v.0.10.0) or Nanopolish (v.0.11.1) [68].

Segregation of reads into parental alleles was performed with WhatsHap (v.0.18) [65], using only new called high-confidence variants.

Acknowledgements

A machine generated summary based on the work of Gilpatrick, Timothy; Lee, Isac; Graham, James E.; Raimondeau, Etienne; Bowen, Rebecca; Heron, Andrew; Downs, Bradley; Sukumar, Saraswati; Sedlazeck, Fritz J; Timp, Winston 2020 in Nature Biotechnology.

Continuous evolution of SpCas9 variants compatible with non-G PAMs.

https://doi.org/10.1038/s41587-020-0412-8

Abstract-Summary

The targeting scope of Streptococcus pyogenes Cas9 (SpCas9) and its engineered variants is largely restricted to protospacer-adjacent motif (PAM) sequences containing G bases.

The targeting capabilities of these evolved variants and SpCas9-NG were characterized in HEK293T cells using a library of 11,776 genomically integrated protospacer–sgRNA pairs containing all possible NNNN PAMs.

These new evolved SpCas9 variants, together with previously reported variants, in principle enable targeting of most NR PAM sequences and substantially reduce the fraction of genomic sites that are inaccessible by Cas9-based methods.

Main

To expand the range of targetable genomic loci, researchers have used naturally occurring Cas9 orthologs with different PAM specificities [69].

Although these efforts have expanded SpCas9's compatibility from NGG PAM sites to most NG PAM sites [70, 71], locations in the genome lacking G bases remain difficult to access.

Base editing is particularly sensitive to Cas9 PAM compatibility: activity for SpCas9-derived base editors is optimal when the PAM is located ~13–17 nucleotides away from the target base [72].

We characterized these three new variants, as well as SpCas9-NG [71], a previously reported engineered SpCas9 that recognizes NG PAMs, on 92 endogenous human genomic target sites and a library of 11,776 integrated target sites.

The new variants reported here, together with previously reported NG PAM-compatible SpCas9 variants, greatly expand the potentially accessible PAM sequence space of SpCas9 to most NR PAMs.

Results

We also tested the indel formation activity of our evolved variants and SpCas9-NG on endogenous sites containing NGN PAMs in HEK293T cells.

These results suggest that, even though the evolved variants can access a broader set of off-target sequences, they have similar or higher overall DNA specificity and on-target activity than SpCas9 on NGG PAM sites.

SpCas9-NRRH, SpCas9-NRCH and SpCas9-NRTH, together with previously reported NG-PAM-compatible SpCas9 variants [70, 71], expand the targeting scope of SpCas9 to most genomic sites containing NR PAMs and greatly increase the fraction of known human pathogenic single-nucleotide polymorphisms (SNPs) that can theoretically be corrected by base editing.

These results demonstrate that the newly evolved SpCas9 variants enable efficient base editing of previously inaccessible pathogenic SNPs using non-G PAMs and highlight the utility of evaluating multiple protospacer and PAM sequences when targeting a particular SNP.

Discussion

Generating SpCas9 variants active on non-G PAMs required improved selection strategies for evolving Cas9:DNA binding, such as increasing the number of target DNA protospacer-PAM binding sites by using an additional PACE-compatible selection circuit that expresses gVI, implementing a dual-AP split-gIII strategy and limiting the total concentration of functional SpCas9 in the host cell through split-intein SpCas9.

The very large number and diversity of sites in this library allowed us to comprehensively profile the editing activity of these proteins using all NNNN PAMs in a genomic context and further illuminated their sequence preferences, for instance demonstrating that our variants displayed a different and complementary fourth-position PAM preference (H) compared to SpCas9-NG (G).

In light of these sequence preferences, we suggest testing the three major variants reported here along with SpCas9-NG when optimizing targeting efficiency on sites with NR PAMs.

Methods

The lagoon was cycled at 1 volume per hour with 10 mM arabinose for 4 h before infection with phage with the SP56 backbone containing residues 574–1368 of

CAA.P1-2 fused to NpuC. Upon infection, lagoon dilution rates were decreased to 0.5 volume per h. The lagoon dilution rate was increased to 1.5 volumes per hour at 96 h, 2.0 volumes per hour at 136 h, and 3.0 volumes per hour at 168 h. The experiment ended at 192 h. P3: Host cells transformed with pTW199b1, pTW221b1 or pTW221b3 and MP6 were maintained in a 40-ml chemostat.

Acknowledgements

A machine generated summary based on the work of Miller, Shannon M.; Wang, Tina; Randolph, Peyton B.; Arbab, Mandana; Shen, Max W.; Huang, Tony P.; Matuszek, Zaneta; Newby, Gregory A.; Rees, Holly A.; Liu, David R. 2020 in Nature Biotechnology.

Evaluation and minimization of Cas9-independent off-target DNA editing by cytosine base editors.

https://doi.org/10.1038/s41587-020-0414-6

Abstract-Summary

We use these assays to identify CBEs with reduced Cas9-independent deamination and validate via whole-genome sequencing that YE1, a narrowed-window CBE variant, displays background levels of Cas9-independent off-target editing.

We engineered YE1 variants that retain the substrate-targeting scope of high-activity CBEs while maintaining minimal Cas9-independent off-target editing.

The suite of CBEs characterized and engineered in this study collectively offer ~10–100-fold lower average Cas9-independent off-target DNA editing while maintaining robust on-target editing at most positions targetable by canonical CBEs, and thus are especially promising for applications in which off-target editing must be minimized.

Main

Similar to other Cas9-directed genome-editing tools, base editors can bind to off-target genomic loci that have high sequence homology to the target protospacer.

To Cas9-dependent off-target base editing, deamination from Cas9-independent binding of a base editor's deaminase domain to DNA represents a distinct type of off-target base editing.

These off-target edits likely arise from the intrinsic DNA affinity of BE3's deaminase domain, independent of the guide RNA-programmed DNA binding of Cas9.

Unlike Cas9-dependent off-target editing, Cas9-independent deamination occurs at different loci between cells, making it difficult to characterize by targeted high-throughput sequencing.

We describe the development of methods to efficiently evaluate the propensity of a base editor to cause Cas9-independent deamination, and the application of these methods to identify and engineer CBE variants that minimize Cas9-independent DNA editing.

Results

To identify base editor variants that exhibit reduced Cas9-independent deamination relative to BE4 in human cells, we evaluated the same panel of 14 deaminase domains (APOBEC1, CDA, AID, APOBEC3A, eA3A, APOBEC3B, APOBEC3G and FERNY; and APOBEC1 mutants YE1, YE2, YEE, EE, R33A and R33A + K34A) [47, 73–84] in the BE4max architecture for their ability to deaminate dSaCas9-induced R-loops in trans.

The results from the rifampin and HSV-TK resistance assays in bacteria, orthogonal R-loop assay in human cells, kinetic assay in vitro and ssDNA deamination assay in human cells are consistent with a model in which CBEs with deaminases that have a low intrinsic catalytic efficiency (k_{cat}/K_m) for cytosine-containing ssDNA substrates exhibit lower Cas9-independent off-target deamination.

Because YE1 and the CBE variants that we assessed to have minimal Cas9-independent off-target activity all exhibit narrowed on-target DNA-editing windows [82] (YE1-BE4, YE2-BE4, YEE-BE4 and EE-BE4), or a specific DNA sequence context requirement [83] (R33A + K34A-BE4), we sought to expand the targeting scope of these CBEs to increase their overall utility.

Discussion

The assays developed and applied in this study enable rapid and cost-effective profiling of base editors for Cas9-independent deamination of DNA in bacteria and mammalian cells, and complement in vivo methods such as those performed by Yang, Gao and their respective coworkers [85, 86].

Given this landscape, and the fact that the 5×10^{-8} per bp mutation rate attributed to Cas9-independent deamination by BE3 in mouse embryos [85] is lower than the observed rate of spontaneous mutation in many mammalian somatic cell types in vivo [87–91], the optimal choice of base editor depends strongly on a given application's on-target sequence context, on-target PAM availability, target tissue type and the extent to which minimizing low levels of Cas9-independent deamination is critical.

Methods

Cells were washed with PBS and then lysed with 150 µl of lysis buffer consisting of 10 mM Tris–HCl (pH 7), 0.05% SDS and 25 µg ml^{-1} Proteinase K (Thermo Fisher Scientific) at 37 °C for 1 h and then heat inactivated at 80 °C for 30 min.

USER enzyme (1.5 U; New England BioLabs) was added to the purified ssDNA and incubated at 37 °C for 1 h. Then, 10 µl of the resulting solution was combined with 10 µl of loading buffer (0.09 M tris(hydroxymethyl)aminomethane, 0.09 M sodium tetraborate, 10 mM EDTA pH 8.0, 10 M urea, 20% sucrose, 0.1% SDS) and loaded on a 10% Tris–borate-EDTA-urea gel (Bio-Rad) that was pre-run in 0.5 × Tris–borate-EDTA buffer for 15 min at 180 V. The cleaved uracil-containing products were resolved from the uncleaved cytosine-containing starting material by electrophoresis for 30 min at 180 V, and the gel was imaged on a GE Typhoon FLA 7000 imager.

Acknowledgements

A machine generated summary based on the work of Doman, Jordan L.; Raguram, Aditya; Newby, Gregory A.; Liu, David R. 2020 in Nature Biotechnology.

3.2 Base Editing and Prime Editing System

Machine generated keywords: base editor, editor, cytosine, apobec, deaminase, efficiently, point mutation, adenine, rat, abe, variant, dna base, cytidine, base editing, cytidine deaminase.

Programmable editing of a target base in genomic DNA without double-stranded DNA cleavage.

https://doi.org/10.1038/nature17946

Main

We envisioned that direct conversion of one DNA base to another at a programmable target locus without requiring DSBs could increase the efficiency of gene correction relative to HDR without introducing an excess of random indels.

We therefore restored the catalytic His residue at position 840 in the Cas9 HNH domain of BE2 (ref 10), resulting in the third-generation base editor (BE3, APOBEC–XTEN–dCas9(A840H)–UGI) that nicks the non-edited strand containing a G opposite the edited U. BE3 retains the Asp10Ala mutation in Cas9 that prevents dsDNA cleavage, and also retains UGI to suppress UDG-initiated BER of the editing strand.

These results collectively establish that base editing can induce much more efficient targeted single-base editing in human cells than Cas9-mediated HDR, and with substantially less (BE3) or almost no (BE2) indel formation.

Methods

To generate the Cy3-labelled dsDNA substrates, the 80-nt strands (5 μl of a 100 μM solution) were combined with the Cy3-labelled primer (5 μl of a 100 μM solution) in NEBuffer 2 (38.25 μl of a 50 mM NaCl, 10 mM Tris–HCl, 10 mM $MgCl_2$, 1 mM DTT, pH 7.9 solution, New England Biolabs) with dNTPs (0.75 μl of a 100 mM solution) and heated to 95 °C for 5 min, followed by a gradual cooling to 45 °C at a rate of 0.1 °C per s. After this annealing period, Klenow exo⁻ (5 U, New England Biolabs) was added and the reaction was incubated at 37 °C for 1 h. The solution was diluted with buffer PB (250 μl, Qiagen) and isopropanol (50 μl) and purified on a QIAprep spin column (Qiagen), eluting with 50 μl of Tris buffer.

Acknowledgements

A machine generated summary based on the work of Komor, Alexis C.; Kim, Yongjoo B.; Packer, Michael S.; Zuris, John A.; Liu, David R. 2016 in Nature.

Increasing the genome-targeting scope and precision of base editing with engineered Cas9-cytidine deaminase fusions.

https://doi.org/10.1038/nbt.3803

Main

We incorporated the corresponding APOBEC1 mutations into BE3 and evaluated their effect on base-editing efficiency and editing window width in HEK293T cells at two C-rich genomic sites containing Cs at positions 3, 4, 5, 6, 8, 9, 10, 12, 13, and 14 (site A); or containing Cs at positions 5, 6, 7, 8, 9, 10, 11, and 13 (site B).

We compared the base-editing outcomes of BE3, YE1-BE3, YE2-BE3, EE-BE3, and YEE-BE3 in HEK293T cells targeting four well-studied human genomic sites that contain multiple Cs within the BE3 activity window [47].

These findings together suggest that the decreased apparent processivity of these narrow-window base editors favors conversion of only a single C at target sites containing multiple Cs within the BE3 editing window.

These developments substantially expand the targeting scope of base editing by developing base editors that use Cas9 variants with different PAM specificities, and by developing a collection of deaminase mutants with varying editing window widths.

Methods

PCR was performed using Q5 Hot Start High-Fidelity DNA Polymerase (New England BioLabs).

DNA vector amplification was carried out using NEB 10beta competent cells (New England BioLabs).

750 ng of BE and 250 ng of sgRNA expression plasmids were transfected using 1.5 μl of Lipofectamine 2000 (ThermoFisher Scientific) per well according to the manufacturer's protocol.

500 ng of BE and 250 ng of sgRNA expression plasmids were transfected into U2OS cells using a Lonza 4D-Nucleofector with the DN-100 program according to the manufacturer's protocols.

Transfected cells were harvested after 3 d. The genomic DNA was isolated using the Agencourt DNAdvance Genomic DNA Isolation Kit (Beckman Coulter) according to the manufacturer's instructions.

PCR amplification was carried out with Phusion hot-start II DNA polymerase (ThermoFisher) according to the manufacturer's instructions.

This program trims sequencing reads to the 20 nucleotide protospacer sequence as determined by a perfect match for the 7 nucleotide sequences that should flank the target site.

Acknowledgements

A machine generated summary based on the work of Kim, Y Bill; Komor, Alexis C; Levy, Jonathan M; Packer, Michael S; Zhao, Kevin T; Liu, David R 2017 in Nature Biotechnology.

Programmable base editing of A•T to G•C in genomic DNA without DNA cleavage.

https://doi.org/10.1038/nature24644

Abstract-Summary

We describe adenine base editors (ABEs) that mediate the conversion of A•T to G•C in genomic DNA.

Extensive directed evolution and protein engineering resulted in seventh-generation ABEs that convert targeted A•T base pairs efficiently to G•C (approximately 50% efficiency in human cells) with high product purity (typically at least 99.9%) and low rates of indels (typically no more than 0.1%).

ABEs introduce point mutations more efficiently and cleanly, and with less off-target genome modification, than a current Cas9 nuclease-based method, and can install disease-correcting or disease-suppressing mutations in human cells.

With previous base editors, ABEs enable the direct, programmable introduction of all four transition mutations without double-stranded DNA cleavage.

Main

The ability to convert A•T base pairs to G•C base pairs at target loci in the genomic DNA of unmodified cells could therefore make it possible to correct a substantial fraction of human disease-associated SNPs.

The most commonly used base editors are third-generation designs (BE3) comprising (i) a catalytically impaired CRISPR–Cas9 mutant that cannot make DSBs; (ii) a single-strand-specific cytidine deaminase that converts C to uracil (U) within an approximately five-nucleotide window in the single-stranded DNA bubble created by Cas9; (iii) a uracil glycosylase inhibitor (UGI) that impedes uracil excision and downstream processes that decrease base editing efficiency and product purity [92]; and (iv) nickase activity to nick the non-edited DNA strand, directing cellular DNA repair processes to replace the G-containing DNA strand [47, 92].

We used protein evolution and engineering to develop a new class of adenine base editors (ABEs) that convert A•T to G•C base pairs in DNA in bacteria and human cells.

Evolution of an adenine deaminase that processes DNA

These results suggest that the inability of these natural adenine deaminase enzymes to accept DNA precludes their direct use in an ABE.

Given these results, we sought to evolve an adenine deaminase that accepts DNA as a substrate.

A•T to G•C conversion at the H193Y mutation should restore chloramphenicol resistance, linking ABE activity to bacterial survival.

Described base editors [47, 51, 92, 93] exploit the use of cytidine deaminase enzymes that operate on single-stranded DNA but reject double-stranded DNA.

These results indicate that mutations at or near TadA D108 enable TadA to perform adenine deamination on DNA substrates.

The TadA A106V and D108N mutations were incorporated into a mammalian codon-optimized TadA–Cas9 nickase fusion construct that replaces dCas9 with the Cas9 D10A nickase used in BE3 to manipulate cellular DNA repair to favour the desired base editing outcomes [47], and adds a C-terminal nuclear localization signal (NLS).

Improved deaminase variants and ABE architectures.

As a final ABE2 engineering study, we investigated the effect of TadA* dimerization on base editing efficiency.

We hypothesized that tethering an additional wild-type or evolved TadA monomer might improve base editing in mammalian cells by minimizing reliance on inter-molecular ABE dimerization.

We determined which of the two TadA* subunits within the TadA*–ABE2.1 fusion was responsible for catalysing conversion of adenine to inosine.

We introduced an inactivating E59A mutation [94] into either the N-terminal or the internal TadA* monomer of ABE2.9.

These results establish that the internal TadA subunit is responsible for catalysing adenine deamination.

ABEs that efficiently edit a subset of targets

We performed a third round of bacterial evolution starting with TadA*2.1–dCas9 to increase editing efficiency further.

The results from six genomic loci with different sequence contexts surrounding the target adenine suggested that ABEs from rounds 1–3 strongly preferred target sequence contexts of YAC, where Y is T or C. This preference is likely to have been inherited from the substrate specificity of native E. coli TadA, which deaminates the adenine in the UAC anticodon of tRNA^{Arg}.

We performed a fifth round of evolution to increase ABE catalytic performance and broaden target sequence compatibility.

In E. coli, endogenous wild-type TadA is provided in trans, potentially explaining the difference between bacterial selection phenotypes and mammalian cell editing efficiencies.

A heterodimeric construct containing wild-type E. coli TadA fused to an internal evolved TadA* (ABE5.3) exhibited greatly improved editing efficiency compared to homodimeric ABE5.1 with two identical evolved TadA* domains.

Highly active ABEs with broad sequence compatibility

During the sixth round of evolution, we aimed to remove any non-beneficial mutations by DNA shuffling and to re-examine mutations from previous rounds of evolution that may benefit ABE performance once liberated from negative epistasis with other mutations.

Characterization of late-stage ABEs

The remarkable product purity of all tested ABE variants suggests that the activity or abundance of enzymes that remove inosine from DNA may be low compared to those of uracil N-glycosylase (UNG), resulting in minimal base excision repair following ABE editing.

We compared the efficiency of ABE7.10-catalysed A•T to G•C editing to that of a current Cas9 nuclease-mediated HDR method, CORRECT [95].

Although HDR is well-suited to introduce insertions and deletions into genomic DNA, these results demonstrate that ABE7.10 can introduce A•T to G•C point mutations with much higher efficiency and far fewer undesired products than a current Cas9 nuclease-mediated HDR method.

As no method yet exists to comprehensively profile the off-target activity of ABEs, we assumed that off-target ABE editing occurred primarily at the off-target sites that are edited when Cas9 nuclease is complexed with the same guide RNA, as is the case with BE3 [47, 93, 96].

Installation of disease-relevant mutations with ABE

We tested the potential of ABEs to introduce disease-suppressing mutations and to correct pathogenic mutations in human cells.

We transfected DNA encoding ABE7.10 and a guide RNA that places the target adenine at protospacer position 5 into an immortalized lymphoblastoid cell line (LCL) harbouring the HFE C282Y genomic mutation.

Although much additional research is needed to develop these and other ABE editing strategies into potential clinical therapies for diseases with a genetic component, including the development of ABEs that accept a wide variety of PAMs [51], these examples demonstrate the potential of ABEs to correct disease-driving mutations, and to install mutations known to suppress genetic disease phenotypes, in human cells.

With BE3 [47] and BE4 [92], these ABEs advance the field of genome editing by enabling the direct installation of all four transition mutations at target loci in living cells with a minimum of undesired byproducts.

Methods

The full-length reassembled product was amplified by PCR with the following conditions: 15 µl unpurified internal assembly was combined with 1 µM each of USER primers NMG-825 and NMG-826, 100 µl Phusion U Green Multiplex PCR Master Mix and H_2O to a final volume of 200 µl, 63 °C annealing, extension time 30 s. The PCR product was purified by gel electrophoresis and assembled using the USER method into the corresponding ecTadA*–XTEN–dCas9 backbone with corresponding flanking USER junctions generated from amplification of the backbone with USER primers NMG-799 and NMG-824 as before.

After 12–14 h, cells were transfected at ~70% confluency with 750 ng Cas9 or base editor plasmid, 250 ng sgRNA expression plasmid, 1.5 μl Lipofectamine 3000 (Thermo Fisher Scientific), and for HDR assays 0.7 μg single-stranded donor DNA template (100 nt, PAGE-purified from IDT) according to the manufacturer's instructions.

Acknowledgements

A machine generated summary based on the work of Gaudelli, Nicole M.; Komor, Alexis C.; Rees, Holly A.; Packer, Michael S.; Badran, Ahmed H.; Bryson, David I.; Liu, David R. 2017 in Nature.

Adenine base editing in an adult mouse model of tyrosinaemia.

https://doi.org/10.1038/s41551-019-0357-8

Abstract-Summary

To traditional CRISPR–Cas9 homology-directed repair, base editing can correct point mutations without supplying a DNA-repair template.

We show in a mouse model of tyrosinaemia that hydrodynamic tail-vein injection of plasmid DNA encoding the adenine base editor (ABE) and a single-guide RNA (sgRNA) can correct an A > G splice-site mutation.

We also generated FAH$^+$ hepatocytes in the liver via lipid-nanoparticle-mediated delivery of a chemically modified sgRNA and an mRNA of a codon-optimized base editor that displayed higher base-editing efficiency than the standard ABEs.

Main

Major caveats of correcting point mutations through CRISPR genome editing include system efficiency, the introduction of double-stranded DNA breaks and the need to provide a DNA-repair template.

Local delivery of ABEs by intramuscular injection of a trans-splicing adeno-associated virus was recently reported [97]; however, systemic delivery of ABEs for correction of liver disease in adult animals has not been investigated.

To explore the therapeutic potential of ABEs in the liver of adult animals, we chose a mouse model of hereditary tyrosinaemia type I (HTI), a fatal genetic disease.

The Fah$^{mut/mut}$ mouse model [98, 99] has a homozygous G•C to A•T point mutation in the last nucleotide of exon 8, resulting in exon skipping and loss of FAH.

We and others recently reported that CRISPR can correct this mutation in Fah through HDR [100–102] or allelic exchange [103].

Following correction by CRISPR, liver cells that express the FAH enzyme, through their selective advantage, expand and repopulate the liver [98].

Results

We tested different doses of two ABE enzymes with different base-editing windows [48], ABE6.3 and ABE7.10, in Fah mutant mouse embryonic fibroblasts (MEFs).

These data indicate that adenine base editing rescues the disease phenotype caused by the Fah mutation in vivo.

To determine whether adenine base editing successfully corrects the HTI Fah splicing mutation in exon 8, we performed reverse transcription PCR (RT–PCR) in liver mRNA using primers that spanned exons 5 and 9.

These data indicate that in vivo delivery of ABEs corrects the G•C to T•A mutation in a subset of liver cells and generates functional exon 8-containing Fah mRNA.

Compared with HDR using short homologous arms (site 1:1.7 $\pm$ 1.1%; site 2: 1.8 $\pm$ 0.8%), ABE6.3 mediated substantially higher A•T to G•C conversion rates at the targeted sites (site 1: 30 $\pm$ 6.3%; site 2: 7.3 $\pm$ 1.6%), whereas RA6.3 further increased base-editing efficiency (site 1: 61 $\pm$ 6.3%; site 2: 34 $\pm$ 3.2%).

Discussion

We codon-optimized ABE6.3 and showed that in vivo delivery of ABEs was able to correct the mutation in Fah in vivo without inducing high level of indels.

As an initial delivery vehicle, our study also showed that non-viral LNP delivery of ABE mRNA and sgRNA generates FAH$^+$ hepatocytes in vivo, albeit with low efficiency.

Methods to improve delivery and to enhance mRNA stability and/or translation will be required to broaden the therapeutic application of ABEs.

Methods

Vectors for hydrodynamic tail-vein injection were prepared using the EndoFreeMaxi kit (Qiagen).

For hydrodynamic liver injection, plasmids suspended in 2 ml saline were injected via the tail vein in 5–7 s into 8–10-week-old Fah$^{mut/mut}$ mice.

1 mg kg^{-1} LNP RA6.3 mRNA and 0.5 mg kg^{-1} LNP Fah SgRNA Were Injected in 8–10-Week-Old Female Fah$^{mut/mut}$ Mice via Tail-Vein Injection.

The mice were injected with 3–4 doses (every 3 days) and kept on NTBC water.

Mice were euthanized 5 days after the last injection and organs were collected for analyses.

Deep-sequencing libraries were made from approximately 1–100 ng of the PCR products.

After 24 h, cells were transfected with 1 μg Cas9 or base editors, 300 ng sgRNA expression plasmid, 6 μl Lipofectamine 3000 (Thermo Fisher Scientific) and for HDR assays, 0.7 μg single-stranded donor DNA template (100 nt, PAGE-purified from IDT) [48].

Acknowledgements

A machine generated summary based on the work of Song, Chun-Qing; Jiang, Tingting; Richter, Michelle; Rhym, Luke H.; Koblan, Luke W.; Zafra, Maria Paz; Schatoff, Emma M.; Doman, Jordan L.; Cao, Yueying; Dow, Lukas E.; Zhu, Lihua

Julie; Anderson, Daniel G; Liu, David R.; Yin, Hao; Xue, Wen 2019 in Nature Biomedical Engineering.

Transcriptome-wide off-target RNA editing induced by CRISPR-guided DNA base editors.

https://doi.org/10.1038/s41586-019-1161-z

Abstract-Summary

The most widely used cytosine base editors (CBEs) induce deamination of DNA cytosines using the rat APOBEC1 enzyme, which is targeted by a linked Cas protein–guide RNA complex [47, 104].

We show that a CBE with rat APOBEC1 can cause extensive transcriptome-wide deamination of RNA cytosines in human cells, inducing tens of thousands of C-to-U edits with frequencies ranging from 0.07% to 100% in 38–58% of expressed genes.

We engineered two CBE variants bearing mutations in rat APOBEC1 that substantially decreased the number of RNA edits (by more than 390-fold and more than 3800-fold) in human cells.

We show that an adenine base editor [48] can also induce transcriptome-wide RNA edits.

These results have implications for the use of base editors in both research and clinical settings, illustrate the feasibility of engineering improved variants with reduced RNA editing activities, and suggest the need to more fully define and characterize the RNA off-target effects of deaminase enzymes in base editor platforms.

Extended:

Investigators were not blinded to experimental conditions or outcome assessments.

Main

To test whether transcriptome-wide RNA editing could also occur in a human cell line from a tissue source other than the liver, we examined BE3 with two gRNAs (targeted to sites in the human RNF2 and EMX1 genes) in human embryonic kidney (HEK293T) cells.

To engineer selective curbing of unwanted RNA editing (SECURE) variants that would show reduced RNA editing but retain efficient on-target DNA base editing, we screened 16 BE3 editors with various APOBEC1 mutations that have previously been reported to reduce RNA C-to-U editing [105–109].

To characterize RNA editing by these two variants more rigorously, we performed RNA-seq experiments using the RNF2 gRNA in transfected HEK293T cells sorted for high expression of wild-type BE3, BE3-R33A, BE3-R33A/K34A or a catalytically impaired BE3-E63Q mutant [106].

These experiments were performed without sorting for GFP expression so that DNA editing activities were assessed without the benefit of higher BE3 variant expression used in the RNA-seq studies described above.

Methods

Cells were then transfected with 37.5 μg base editor or negative control (nCas9(D10A)-UGI-NLS(SV40) or bpNLS-32AA linker-nCas9(D10A)-bpNLS) plasmid fused to P2A-eGFP, 12.5 μg gRNA expression plasmid, and 150 μl TransIT-293 (for HEK293T, Mirus) or transfeX (for HepG2, ATCC) according to the manufacturer's protocols.

For experiments to validate the DNA on-target activity of SECURE variants, 1.5×10^4 HEK293T cells were seeded into 96-well flat-bottom cell-culture plates (Corning) and transfected 24 h after seeding with 220 ng DNA (165 ng base editor or negative control plasmid and 55 ng gRNA expression plasmid) and 0.66 μl TransIT-293.

For scatter plots, the background rates of C-to-T or A-to-G alterations in the control sample were subtracted from base editor-treated sample to compute the DNA editing rate attributable to the base editor; in these same scatter plots, note that we only call RNA edits in base-editor-treated samples that do not appear in their corresponding control samples (nCas9–UGI–NLS for CBE or NLS–nCas9–NLS for ABE) as processed by our filtering pipeline (see 'RNA sequence variant filtering methods') and thus background rates of RNA editing are already accounted for in the depiction of these data.

Acknowledgements

A machine generated summary based on the work of Grünewald, Julian; Zhou, Ronghao; Garcia, Sara P.; Iyer, Sowmya; Lareau, Caleb A.; Aryee, Martin J.; Joung, J. Keith 2019 in Nature.

Search-and-replace genome editing without double-strand breaks or donor DNA.

https://doi.org/10.1038/s41586-019-1711-4

Abstract-Summary

We describe prime editing, a versatile and precise genome editing method that directly writes new genetic information into a specified DNA site using a catalytically impaired Cas9 endonuclease fused to an engineered reverse transcriptase, programmed with a prime editing guide RNA (pegRNA) that both specifies the target site and encodes the desired edit.

We used prime editing in human cells to correct, efficiently and with few byproducts, the primary genetic causes of sickle cell disease (requiring a transversion in HBB) and Tay–Sachs disease (requiring a deletion in HEXA); to install a protective transversion in PRNP; and to insert various tags and epitopes precisely into target loci.

Prime editing shows higher or similar efficiency and fewer byproducts than homology-directed repair, has complementary strengths and weaknesses compared to base editing, and induces much lower off-target editing than Cas9 nuclease at known Cas9 off-target sites.

Prime editing substantially expands the scope and capabilities of genome editing, and in principle could correct up to 89% of known genetic variants associated with human diseases.

Main

Base editing can efficiently install the four transition mutations (C → T, G → A, A → G, and T → C) without requiring DSBs in many cell types and organisms, including mammals [47, 72, 110, 111], but cannot currently perform the eight transversion mutations (C → A, C → G, G → C, G → T, A → C, A → T, T → A, and T → G), such as the T•A-to-A•T mutation needed to directly correct the most common cause of sickle cell disease (HBB(E6V)).

We describe the development of prime editing, a 'search-and-replace' genome editing technology that mediates targeted insertions, deletions, all 12 possible base-to-base conversions, and combinations thereof in human cells without requiring DSBs or donor DNA templates.

By enabling precise targeted insertions, deletions, and all 12 possible classes of point mutations without requiring DSBs or donor DNA templates, prime editing has the potential to advance the study and correction of the vast majority of pathogenic alleles.

Prime editing strategy

We envisioned the generation of guide RNAs that both specify the DNA target and contain new genetic information that replaces target DNA nucleotides.

Although hybridization of the perfectly complementary 5′ flap to the unedited strand is likely to be thermodynamically favoured, 5′ flaps are the preferred substrate for structure-specific endonucleases such as FEN1 [112], which excises 5′ flaps generated during lagging-strand DNA synthesis and long-patch base excision repair.

Validation in vitro and in yeast

We tested whether the 3′ end of the protospacer-adjacent motif (PAM)-containing DNA strand cleaved by the RuvC nuclease domain of Cas9 was sufficiently accessible to prime reverse transcription.

These results demonstrate that nicked DNA exposed by dCas9 is competent to prime reverse transcription from a pegRNA.

These experiments establish that 3′-extended pegRNAs can direct Cas9 nickase and template reverse transcription in vitro.

Prime editor 1

These results suggest that wild-type M-MLV RT fused to Cas9 requires longer PBS sequences for genome editing in human cells compared to what is required in vitro using the commercial variant of M-MLV RT supplied in trans.

We designated this M-MLV RT fused to the C terminus of Cas9(H840A) nickase as PE1.

Prime editor 2

We hypothesized that engineering the RT in PE1 might improve the efficiency of DNA synthesis during prime editing.

We constructed 19 variants of PE1 containing a variety of RT mutations to evaluate their editing efficiency in human cells.

Optimization of pegRNAs

No PBS length or G/C content level was strictly predictive of editing efficiency, suggesting that other factors such as DNA primer or RT template secondary structure also influence editing activity.

Because many RT template lengths support prime editing, we recommend designing pegRNAs so that the first base of the 3' extension is not C.

Prime editor 3 systems

As the edited DNA strand is also nicked to initiate prime editing, we tested a variety of nick locations on the non-edited strand to minimize DSBs that lead to indels.

We recommend starting with non-edited strand nicks about 50 bp from the pegRNA-mediated nick, and testing alternative nick locations if indel frequencies exceed acceptable levels.

When it is possible to nick the non-edited strand with an sgRNA that requires editing before nicking, the PE3b system offers PE3-like editing levels while greatly reducing indel formation.

As an NGG PAM on either DNA strand occurs on average every 8 bp, far less than edit-to-PAM distances that support efficient prime editing, prime editing is not substantially constrained by the availability of a nearby PAM sequence, in contrast to other precision editing methods [72, 95, 113].

We observed very high editing efficiencies (52–78%) for precise 5-, 10-, 15-, 25-, and 80-bp deletions, with indels averaging $11 \pm 4.8\%$.

Prime editing compared with base editing

We compared PEs and CBEs at three genomic loci that contain multiple target cytosines in the canonical base editing window (protospacer positions 4–8, counting the PAM as positions 21–23) using current-generation CBEs [114] without or with nickase activity (BE2max and BE4max, respectively), or using analogous PE2 and PE3 prime editing systems.

For the installation of precise edits (with no bystander editing), the efficiency of prime editing greatly exceeded that of base editing at the above sites, which, like most genomic DNA sites, contain multiple cytosines within the base editing window.

When a single target nucleotide is present within the base editing window, or when bystander edits are acceptable, current base editors are typically more efficient and generate fewer indels than prime editors.

When multiple cytosines or adenines are present and bystander edits are undesirable, or when PAMs that position target nucleotides for base editing are not available, prime editors offer substantial advantages.

Off-target prime editing

Prime editing requires target DNA–pegRNA spacer complementarity for the Cas9 domain to bind, target DNA–pegRNA PBS complementarity to initiate pegRNA-templated reverse transcription, and target DNA–RT product complementarity for flap resolution.

To test whether these three distinct DNA hybridization steps reduce off-target prime editing compared to editing methods that require only target–guide RNA complementarity, we treated HEK293T cells with PE3 or PE2 and 16 pegRNAs that target four genomic loci, each of which has at least four well-characterized Cas9 off-target sites [19, 115].

At the HEK4 off-target 3 site that was edited by Cas9 with pegRNA1 at 97% efficiency, PE2 with pegRNA1 resulted in only 0.2% off-target editing despite sharing the same pegRNA, demonstrating how the two additional hybridization events required for prime editing can greatly reduce off-target modification.

Prime editing pathogenic mutations

We tested the ability of PE3 to directly install or correct in human cells transversion, insertion, and deletion mutations that cause genetic diseases.

These results establish the ability of prime editing in human cells to install or correct transversion, insertion, or deletion mutations that cause or confer resistance to disease efficiently, and with few byproducts.

Other cell lines and primary neurons

These data indicate that cell lines other than HEK293T support prime editing, although editing efficiencies vary by cell type and are generally less efficient than in HEK293T cells.

Prime editing compared with HDR

Cas9-initiated HDR in all cases successfully installed the desired edit, but with far higher levels of indel byproducts than with PE3, as expected given that Cas9 induces DSBs.

Discussion and future directions

The ability to insert arbitrary DNA sequences with single-nucleotide precision is an especially promising capability of prime editing.

The prime editing experiments described here performed 19 insertions up to 44 bp, 23 deletions up to 80 bp, 119 point mutations including 83 transversions, and 18 combination edits at 12 endogenous loci in the human and mouse genomes at locations ranging from 3 bp upstream to 29 bp downstream of a PAM without making explicit DSBs.

By enabling precise targeted transitions, transversions, insertions, and deletions in the genomes of mammalian cells without requiring DSBs, donor DNA templates, or HDR, however, prime editing provides a new search-and-replace capability that substantially expands the scope of genome editing.

Methods

Between 16 and 24 h after seeding, cells were transfected at approximately 60% confluency with 1 µl lipofectamine 2000 (Thermo Fisher Scientific) according to the manufacturer's protocols and 750 ng PE plasmid, 250 ng pegRNA plasmid, and 83 ng sgRNA plasmid (for PE3 and PE3b).

Cells were cultured for 3 days following transfection, after which the medium was removed, the cells were washed with 1 × PBS solution (Thermo Fisher Scientific), and genomic DNA was extracted by the addition of 150 µl of freshly prepared lysis buffer (10 mM Tris–HCl, pH 7.5; 0.05% SDS; 25 µg/ml proteinase K (ThermoFisher Scientific)) directly into each well of the tissue culture plate.

Between 16 and 24 h after seeding, cells were transfected at approximately 70% confluency with 1 µl lipofectamine 2000 (Thermo Fisher Scientific) according to the manufacturer's protocols and 750 ng PE2-P2A-GFP plasmid, 250 ng pegRNA plasmid, and 83 ng sgRNA plasmid.

Acknowledgements

A machine generated summary based on the work of Anzalone, Andrew V.; Randolph, Peyton B.; Davis, Jessie R.; Sousa, Alexander A.; Koblan, Luke W.; Levy, Jonathan M.; Chen, Peter J.; Wilson, Christopher; Newby, Gregory A.; Raguram, Aditya; Liu, David R. 2019 in Nature.

CRISPR C-to-G base editors for inducing targeted DNA transversions in human cells.

https://doi.org/10.1038/s41587-020-0609-x

Abstract-Summary

We describe the engineering of two base editor architectures that can efficiently induce targeted C-to-G base transversions, with reduced levels of unwanted C-to-W (W = A or T) and indel mutations.

One of these C-to-G base editors (CGBE1), consists of an RNA-guided Cas9 nickase, an Escherichia coli–derived uracil DNA N-glycosylase (eUNG) and a rat APOBEC1 cytidine deaminase variant (R33A) previously shown to have reduced off-target RNA and DNA editing activities [116, 117].

CGBE1 and miniCGBE1 enable C-to-G edits and will serve as a basis for optimizing C-to-G base editors for research and therapeutic applications.

Main

Given this observation about ABE-mediated C-to-G alterations, we wondered whether we could induce these edits more efficiently by modifying the BE4max cytosine base editor (CBE) [104, 114], which harbors an enzyme actually intended to deaminate cytosines (the rat APOBEC1 cytidine deaminase).

Given its higher C-to-G editing activity, we chose the eUNG-BE4max(R33A)ΔUGI fusion (hereafter referred to as C-to-G Base Editor 1 (CGBE1)) for additional characterization.

Among four target sites we tested, we also found that our CGBE variants generally induced higher frequencies of desired C-to-G edits than PE2 or PE3 prime editors directed to induce the same alterations.

A better understanding of the mechanistic parameters that govern both the frequencies and the product purity of C-to-G edits may suggest additional strategies to further increase the targeting range and efficiency of CGBEs.

Methods

K562 cells were electroporated using the SF Cell Line Nucleofector X Kit (Lonza), according to the manufacturer's protocol with 2×10^5 cells per nucleofection and 800 ng control or base/prime editor plasmid, 200 ng gRNA or pegRNA plasmid and 83 ng nicking gRNA plasmid (for PE3).

U2OS cells were electroporated using the SE Cell Line Nucleofector X Kit (Lonza) with 2×10^5 cells and 800 ng control or base/prime editor plasmid, 200 ng gRNA or pegRNA and 83 ng nicking gRNA (for PE3).

HeLa cells were electroporated using the SE Cell Line 4D-Nucleofector X Kit (Lonza) with 5×10^5 cells and 800 ng control or base/prime editor, 200 ng gRNA or pegRNA and 83 ng nicking gRNA (for PE3).

Acknowledgements

A machine generated summary based on the work of Kurt, Ibrahim C.; Zhou, Ronghao; Iyer, Sowmya; Garcia, Sara P.; Miller, Bret R.; Langner, Lukas M.; Grünewald, Julian; Joung, J. Keith 2020 in Nature Biotechnology.

3.3 Optimization of Guide RNA Design

Machine generated keywords: predict, algorithm, prediction, grna, mismatch, offtarget, offtarget activity, ontarget, optimal, knockdown, rule, design, hit, profile, critical.

Optimized sgRNA design to maximize activity and minimize off-target effects of CRISPR-Cas9.

https://doi.org/10.1038/nbt.3437

Main

Initial libraries were designed with little knowledge of sgRNA activity rules, a critical design parameter, as interpreting screening data requires consistency among multiple sgRNAs targeting the same gene to distinguish true hits from false positives.

We present the design and characterization of human and mouse genome-wide sgRNA libraries based on our previously published rules for predicting on-target efficiency [118].

Building on screening data generated with the new libraries and large-scale assessment of off-target activity, we develop improved algorithms for on- and off-target activity prediction, allowing further optimization of our genome-wide libraries.

Results

For all sites with the canonical NGG PAM, in addition to the perfect-match sgRNAs, we introduced three types of sgRNA mutations: first, all 1-nucleotide deletions; second, all 1-nucleotide insertions; third, all 1-nucleotide mismatches to the target DNA, generating a library with 27,897 unique sgRNAs.

We tested the ability of the CFD score to predict off-target, imperfect-match activity for an independent data set of 89 sgRNAs designed to target H2-D (also known as,) some of which were previously shown to produce effective protein knockout of H2-K (also known as), a gene with highly similar sequence [118].

The Avana library contains 4950 sgRNAs targeting these genes, and we examined their behavior in the viability screen in A375 cells described above.

This observation suggests that sgRNAs with more predicted off-target sites are more frequently erroneously depleted in a negative selection screen, and thus avoidance of such promiscuous sgRNAs will lead to improved library performance.

Discussion

Predictions of sgRNA activity with Rule Set 1 were imperfect, however, leaving opportunity for further improvements in library design.

By doubling the size of the sgRNA activity data set and identifying a more effective modeling approach, we developed Rule Set 2, which shows demonstrably improved performance versus Rule Set 1 across multiple data sets from multiple laboratories.

SgRNA designs based on Rule Set 2 (such as our Brunello and Brie libraries) should deliver substantial additional improvements in library performance.

Whereas Rule Sets 1 and 2 represent data-driven, quantitative models of on-target sgRNA effectiveness, a quantitative analysis of off-target interactions based on similarly large-scale data has been lacking, and the ability to predict off-target effects therefore quite limited.

Smaller libraries with a high fraction of active and specific sgRNAs, enabled by better on- and off-target activity predictions, will facilitate cost-effective screening across a range of model systems.

Methods

The following list gives final distributions of sgRNAs chosen for inclusion in Avana and Asiago libraries per tier, respectively: i; 57%; 57% ii; 43%; 43% iii; 0.06%; 0.04% iv; 0.02%; 0.01% Criterion B: to mitigate off-target effects, sequence uniqueness of various lengths, counting from the 3′ end of the sgRNA, for example, of all possible sgRNAs targeting protein coding genes, the PAM-proximal (i) 13 nts are unique, (ii) 17 nts are unique, (iii) 20 nts are unique or (iv) the sgRNA sequence is not unique.

MOLM13-Cas9 cells were infected in one biological replicate at 1000 cells per sgRNA, and were selected with puromycin in complete media supplemented with 1% penicillin/streptomycin for the length of the screen.

Acknowledgements

A machine generated summary based on the work of Doench, John G; Fusi, Nicolo; Sullender, Meagan; Hegde, Mudra; Vaimberg, Emma W; Donovan, Katherine F; Smith, Ian; Tothova, Zuzana; Wilen, Craig; Orchard, Robert; Virgin, Herbert W; Listgarten, Jennifer; Root, David E 2016 in Nature Biotechnology.

Evaluation of off-target and on-target scoring algorithms and integration into the guide RNA selection tool CRISPOR.

https://doi.org/10.1186/s13059-016-1012-2

Abstract-Summary

We conduct the first independent evaluation of CRISPR/Cas9 predictions.

We collect data from eight SpCas9 off-target studies and compare them with the sites predicted by popular algorithms.

We identify problems in one implementation but found that sequence-based off-target predictions are very reliable, identifying most off-targets with mutation rates superior to 0.1%, while the number of false positives can be largely reduced with a cutoff on the off-target score.

We also evaluate on-target efficiency prediction algorithms against available datasets.

With novel data from our labs, we find that the optimal on-target efficiency prediction model strongly depends on whether the guide RNA is expressed from a U6 promoter or transcribed in vitro.

To make these guidelines easily accessible to anyone planning a CRISPR genome editing experiment, we built a new website (http://crispor.org) that predicts off-targets and helps select and clone efficient guide sequences for more than 120 genomes using different Cas9 proteins and the eight efficiency scoring systems evaluated here.

Extended:

We hope that the results and resources presented here will aid with future improvements and wider adoption of CRISPR/Cas9 off- and on-target prediction algorithms and reduce the time spent on screening for off-targets and efficient guide sequences.

Background

As more researchers face the task of selecting an optimal Cas9 guide RNA sequence that targets a genome sequence of interest, the overall specificity of the technique is still under discussion: high-throughput cell culture studies have found numerous off-targets not predicted by existing algorithms, sometimes even involving 1-bp indels ("bulges") in the alignment with the guide sequence [19, 119], while studies in

Drosophila, Caenorhabditis elegans, zebrafish, and mice have found virtually no off-target effects [120, 121].

Although published tools and scoring systems allow ranking sequences by specificity [122, 123] and efficiency [118, 119, 124, 125], they are usually limited to a handful of genomes and only few evidence-based recommendations exist to optimize off-target search parameters and on-target efficiency.

Results and discussion

In order to rank potential off-targets, many prediction tools calculate a score based on the position of the mismatches to the guide sequence.

The more recent CFD score [126] is based on the biggest dataset to date, cleavage data obtained by infecting cells with a lentiviral library containing thousands of guides targeting the CD33 gene for all PAMs, including guides for all possible nucleotide mismatches and 1-bp indels at all positions.

The MIT scores of all potential off-targets of a guide can be summarized into the "guide specificity score" defined by [127], which ranges from 0 to 100 (100 = best).

While CRISPOR shows the MIT specificity score as an indicator of guide quality, all potential off-targets are annotated and shown for detailed inspection.

The dataset "Eschstruth" is very small and includes several guides that were selected based on very high Doench scores.

Conclusions

For the remaining sites, sequence-based prediction performance has to be seen relative to the sensitivity of the experimental system used to validate the off-targets.

When using a cutoff on the CFD score, predictions contain 98% of off-target sites validated by whole-genome assays (sensitivities > ~ 0.1%), with a 43% false positive rate.

We hope that the results and resources presented here will aid with future improvements and wider adoption of CRISPR/Cas9 off- and on-target prediction algorithms and reduce the time spent on screening for off-targets and efficient guide sequences.

Methods

The complete dataset consists of 30 guide sequences tested by 36 assays, 634 off-target sequences, and 697 cleavage frequencies, as some off-targets were detected by different assays.

After removal of the two GC-rich guides and 0.01% modification frequency filtering, the filtered dataset contained 225 modification frequency measurements of 179 off-target sequences for 31 tested guide sequences, of which 26 guide sequences contain off-targets >0.01%.

Cleavage efficiency was measured by extracting genomic DNA from around 20 embryos, PCR of the target regions, cloning the result into a TOPO-vector, and shipping for Sanger sequencing a number of colonies in the range 10–20.

We hope that authors of new efficiency scores add their code to this module for easier evaluation and integration into future guide selector websites.

Based on all off-target scores for a guide, a specificity score is calculated using the same formula as on the CRISPR Design website (http://crispr.mit.edu).

Acknowledgements

A machine generated summary based on the work of Haeussler, Maximilian; Schönig, Kai; Eckert, Hélène; Eschstruth, Alexis; Mianné, Joffrey; Renaud, Jean-Baptiste; Schneider-Maunoury, Sylvie; Shkumatava, Alena; Teboul, Lydia; Kent, Jim; Joly, Jean-Stephane; Concordet, Jean-Paul 2016 in Genome Biology.

Predicting the mutations generated by repair of Cas9-induced double-strand breaks.

https://doi.org/10.1038/nbt.4317

Main

Cas9 then cuts DNA at that location, and when the double-strand break is repaired by cellular machinery, frameshift mutations can occur, disabling translation of the correct protein.

Cas9-generated mutations result from imperfect action of DNA repair pathways that are activated to remedy the double-strand break.

Although DNA repair pathways and their key components have been characterized, the biases that favor one mutation over another are not fully understood, especially for the breaks inflicted by Cas9.

The largest current dataset of genomic repair profiles comprises 436 profiles examining 96 unique gRNA sequences using the Cas9 protein from Streptococcuspyogenes [128], recently followed up with studies of more target sites [129, 130].

We present a large-scale measurement of Cas9-generated gRNA repair profiles.

We synthesized over 40,000 DNA constructs, each containing both a gRNA and its target; introduced them into Cas9-expressing cell lines; and sequenced the targeted loci.

Results

Our construct only contains 79 nt of local context owing to limitations of oligonucleotide synthesis yet produces very similar outcomes for 94% of measured cases with sufficient reads (67 of 71), this result confirms that sequence surrounding the cut site is the main determinant of Cas9-induced mutational outcomes.

Insertions, single and double nucleotide deletions, and microhomology-mediated deletions can all be present at frequencies ranging from near 0 to over 50% depending on the target, further highlighting the sequence-specific nature of the repair process.

Microhomology is known to bias repair of Cas9-induced double-strand breaks [131, 132], we first systematically evaluated repair outcomes of targets with different microhomology spans (3–15 nt) and separating distances (0–20 nt).

To test this hypothesis, we developed a computational predictor of the mutational outcomes of a given gRNA, which we call FORECasT (favored outcomes of repair events at Cas9 targets).

Discussion

The Cas9-generated alleles show strong sequence-dependent biases that are reproducible and predictable for dominant categories of mutation (single base insertions, small deletions and microhomology-mediated deletions), despite some variability between genetic backgrounds and species.

Preference for microhomology-mediated repair in stem cells may be linked to increased rates of homology-directed repair, which shares the initial resection step [133], whereas favoring of single base insertions in CHO and RPE-1 lines indicates elevated canonical end-joining activity.

The higher incidence of large insertions in stem cells could similarly be explained by aberrant homology-directed repair, wherein strand invasion occurs in the wrong place, such that DNA synthesis before strand displacement leads to additional sequence being inserted.

Favoring of thymine insertion by this event could indicate either a preference of the DNA repair enzymes (especially polymerases), difference in availability of the required nucleotide triphosphate for incorporation, or propensity of Cas9 to make a staggered rather than blunt cut when thymine is present.

Methods

6218 gRNA–target pairs, all of which were already included in one of the first three subsets above (77 Endogenous, 3777 Genomic (distinct set from 3777 used for across-cell-line comparisons), 2364 Explorative), were ordered as separate oligonucleotides in the purchased pool and were independently cloned with the alternative conventional gRNA scaffold [9].

Once reads were assigned to oligonucleotides, we checked each sample to ensure that the per-oligonucleotide $\log_2$(read count) values (including those both with and without indels) of the Explorative gRNA–target set (since these have no direct targets in the genome) were well correlated (Pearson's R > 0.95, computed using scipy.stats.pearsonr [134]) with those in the original plasmid library to minimize distortion in the measured mutational profiles that could be due to reasons other than Cas9 cutting and subsequent cellular repair.

Acknowledgements

A machine generated summary based on the work of Allen, Felicity; Crepaldi, Luca; Alsinet, Clara; Strong, Alexander J.; Kleshchevnikov, Vitalii; De Angeli, Pietro; Páleníková, Petra; Khodak, Anton; Kiselev, Vladimir; Kosicki, Michael; Bassett, Andrew R.; Harding, Heather; Galanty, Yaron; Muñoz-Martínez, Francisco; Metzakopian, Emmanouil; Jackson, Stephen P.; Parts, Leopold 2018 in Nature Biotechnology.

Increasing the specificity of CRISPR systems with engineered RNA secondary structures.

https://doi.org/10.1038/s41587-019-0095-1

Abstract-Summary

We show that engineering a hairpin secondary structure onto the spacer region of single guide RNAs (hp-sgRNAs) can increase specificity by several orders of magnitude when combined with various CRISPR effectors.

We first demonstrate that designed hp-sgRNAs can tune the activity of a transactivator based on Cas9 from Streptococcus pyogenes (SpCas9).

We then show that hp-sgRNAs increase the specificity of gene editing using five different Cas9 or Cas12a variants.

Our results demonstrate that RNA secondary structure is a fundamental parameter that can tune the activity of diverse CRISPR systems.

Main

The identification and characterization of class 2 CRISPR systems is thus an active area of research, with the overarching goal of finding Cas effectors with novel or improved properties [135–137].

Since the initial characterization of SpCas9, the number of Cas effectors active in mammalian cells has expanded to include compact Cas9 effectors from the type II CRISPR systems, Cas12a (previously Cpf1) effectors with $(A + T)$-rich PAMs from type V systems and RNA-targeting Cas13 variants [44, 138–143].

Methods to increase the specificity of class 2 CRISPR systems through rational design have largely focused on SpCas9 and have adopted two general strategies.

There is a need for a simple method for increasing specificity of diverse CRISPR systems.

Employing rational design and adopting the second strategy, we hypothesized that engineering the sgRNA might serve as a means to regulate diverse CRISPR systems.

Results

Although we ascribe the changes in gene activation to modulation of R-loop formation by hp-sgRNAs, previous studies showed by northern blot that 5' extensions to sgRNAs were efficiently processed to 20-nucleotide spacers [23, 144].

The differences in behavior between hp-sgRNAs and ns-sgRNAs indicate that the secondary structure of the spacer is a critical determinant of CRISPR activity.

To assess the effects of engineered hp-sgRNAs on the nuclease activity and specificity of Cas9 in human cells, we chose spacers that have large numbers of well-characterized off-target sites [19].

Target recognition by Cas12a and R-loop formation mechanisms are also reversed when comparing with that of Cas9: the PAM sequence is located at the 5' end of the target sequence and R-loop formation of the target strand proceeds 3' to 5'. Despite these many differences, we hypothesized that the activity of Cas12a nucleases could also be regulated by spacer secondary structure.

Discussion

Despite the widely ranging biochemical properties of each Cas effector used, we observe consistent behavior of hp-sgRNAs, where CRISPR activity is inhibited as a function of the stability of the secondary structure.

While we do not directly determine the mechanism of hp-sgRNA-driven specificity increase, we hypothesize that it occurs through inhibition of R-loop kinetics, which inhibits the structural transitions of the CRISPR endonuclease that are necessary for activity at off-target sites [145].

These points suggest that sgRNA–endonuclease complex levels are maintained and that observed specificity increases are caused by secondary-structure mediated inhibition of R-loop formation, limiting the conformation change to an activated endonuclease at off-target sites.

Our study considers R-loop formation as the central process governing CRISPR nuclease activity: its modulation allows for more specific genome editing and its modeling facilitates predictions of CRISPR activity.

Methods

The resulting PCR products were purified using Agencourt AMPure beads (Beckman coulter), quantified using Qubit Fluorimeter (Thermo Fisher), pooled and sequenced with 150-base pair (bp) paired-end reads on an Illumina MiSeq instrument.

At each state m, the sgRNA is assumed to be in quasi-equilibrium with the DNA, such that at perfectly matched spacer sites the forward rate (rate of additional guide RNA invasion; m to m + 1) v_f is estimated using the symmetric approximation to be $\exp(-(\Delta G°(m + 1)_{RNA:DNA} - \Delta G°(m + 1)_{DNA:DNA} - \Delta G°(m + 1)_{RNA,SS})/2RT)$, where R is Boltzmann's constant, T is the temperature (here 37 °C to correspond with the parameter set we used) and the 1/2 corrective term is included to satisfy detailed balance.

Acknowledgements

A machine generated summary based on the work of Kocak, D. Dewran; Josephs, Eric A.; Bhandarkar, Vidit; Adkar, Shaunak S.; Kwon, Jennifer B.; Gersbach, Charles A. 2019 in Nature Biotechnology.

Massively parallel Cas13 screens reveal principles for guide RNA design.

https://doi.org/10.1038/s41587-020-0456-9

Abstract-Summary

Type VI CRISPR enzymes are RNA-targeting proteins with nuclease activity that enable specific and robust target gene knockdown without altering the genome.

To define rules for the design of Cas13d guide RNAs (gRNAs), we conducted massively parallel screens targeting messenger RNAs (mRNAs) of a green fluorescent protein transgene, and CD46, CD55 and CD71 cell-surface proteins in human cells.

We measured the activity of 24,460 gRNAs with and without mismatches relative to the target sequences.

We developed a computational model to identify optimal gRNAs and confirm their generalizability, testing 3979 guides targeting mRNAs of 48 endogenous genes.

Extended:

We designed 6967 GFP-targeting guides and added 533 nontargeting guides (NT set) of the same length from randomly generated sequences that did not align to the human genome (hg19) with fewer than 3 mismatches.

Main

To show that our model is generalizable, we designed gRNAs to target the endogenous transcripts of CD46 and CD71, which encode cell-surface proteins, and measured the gRNA knockdown efficacy by flow cytometry.

To increase throughput and test gRNA efficacy predictions for more genes, we first generated a small crRNA library targeting ten essential and ten control genes with both three high-scoring and three low-scoring gRNAs, and monitored their depletion in a gene essentiality screen over time.

Our predictive on-target model based on the GFP-tiling screen could largely separate gRNAs with low knockdown efficacy from those with high efficacy.

We show that our updated on-target model, $RF_{combined}$, can predict Cas13d gRNA target knockdown efficacies, separating poorly performing gRNAs from gRNAs with high efficacy, and generalizing across numerous targets.

We show that crRNA features and target RNA context constrain target knockdown efficacy and, using these data, we developed a model to predict gRNAs with high efficacy.

Methods

From the set of remaining PM gRNA predictions, we manually selected three high-scoring and three low-scoring guides for the HEK293FT cell line screen, to ensure that each guide fell into nonoverlapping regions of the target transcripts.

To compute the null distribution, we calculated the median predicted guides scores of randomly selected gRNAs across 1000 samplings for each n. For the leave-one-out cross-validation we trained on all data from three tiling screens and performed Spearman's rank correlation for the predicted guide efficiency of the held-out fourth screen to the observed $\log_2(FC)$ enrichments.

For both essentiality screens we used ten essential genes (all in HEK293FT and the ten most depleted in A375 cells) and correlated the predicted guide scores from both models to the observed $\log_2(FC)$ guide depletion scores (normalized to 0–100% per gene) of all detected gRNAs (HEK293FT: n = 60 with 6 guides per gene; A375: n = 398 with up to 40 guides per gene).

Acknowledgements

A machine generated summary based on the work of Wessels, Hans-Hermann; Méndez-Mancilla, Alejandro; Guo, Xinyi; Legut, Mateusz; Daniloski, Zharko; Sanjana, Neville E. 2020 in Nature Biotechnology.

References

1. Snowden, A.W., P.D. Gregory, C.C. Case, and C.O. Pabo. 2002. Gene-specific targeting of H3K9 methylation is sufficient for initiating repression in vivo. *Current Biology* 12: 2159–2166.
2. Maeder, M.L., et al. 2013. Targeted DNA demethylation and activation of endogenous genes using programmable TALE-TET1 fusion proteins. *Nature Biotechnology* 31: 1137–1142.
3. Mendenhall, E.M., et al. 2013. Locus-specific editing of histone modifications at endogenous enhancers. *Nature Biotechnology* 31: 1133–1136.
4. Rivenbark, A.G., et al. 2012. Epigenetic reprogramming of cancer cells via targeted DNA methylation. *Epigenetics* 7: 350–360.
5. Konermann, S., et al. 2013. Optical control of mammalian endogenous transcription and epigenetic states. *Nature* 500: 472–476.
6. Keung, A.J., C.J. Bashor, S. Kiriakov, J.J. Collins, and A.S. Khalil. 2014. Using targeted chromatin regulators to engineer combinatorial and spatial transcriptional regulation. *Cell* 158: 110–120.
7. Hsu, P.D., E.S. Lander, and F. Zhang. 2014. Development and applications of CRISPR–Cas9 for genome engineering. *Cell* 157: 1262–1278.
8. Doudna, J.A., and E. Charpentier. 2014. Genome editing. The new frontier of genome engineering with CRISPR-Cas9. *Science* 346: 1258096.
9. Mali, P., et al. 2013. RNA-guided human genome engineering via Cas9. *Science* 339: 823–826.
10. Jinek, M., et al. 2012. A programmable dual-RNA-guided DNA endonuclease in adaptive bacterial immunity. *Science* 337: 816–821.
11. Cong, L., et al. 2013. Multiplex genome engineering using CRISPR/Cas systems. *Science* 339: 819–823.
12. Gao, X. et al. 2014. Comparison of TALE designer transcription factors and the CRISPR/dCas9 in regulation of gene expression by targeting enhancers. *Nucleic Acids Research* 42: e155.
13. Gao, X., et al. 2013. Reprogramming to pluripotency using designer TALE transcription factors targeting enhancers. *Stem Cell Reports* 1: 183–197.
14. Ji, Q., et al. 2014. Engineered zinc-finger transcription factors activate OCT4 (POU5F1), SOX2, KLF4, c-MYC (MYC) and miR302/367. *Nucleic Acids Research* 42: 6158–6167.
15. Deng, W., et al. 2014. Reactivation of developmentally silenced globin genes by forced chromatin looping. *Cell* 158: 849–860.
16. Rada-Iglesias, A., et al. 2011. A unique chromatin signature uncovers early developmental enhancers in humans. *Nature* 470: 279–283.
17. Visel, A., et al. 2009. ChIP-seq accurately predicts tissue-specific activity of enhancers. *Nature* 457: 854–858.
18. Gibson, D.G., et al. 2009. Enzymatic assembly of DNA molecules up to several hundred kilobases. *Nature Methods* 6: 343–345.
19. Tsai, S.Q., et al. 2015. GUIDE-seq enables genome-wide profiling of off-target cleavage by CRISPR-Cas nucleases. *Nature Biotechnology* 33: 187–197.
20. Fu, Y., J.D. Sander, D. Reyon, V.M. Cascio, and J.K. Joung. 2014. Improving CRISPR-Cas nuclease specificity using truncated guide RNAs. *Nature Biotechnology* 32: 279–284.

21. Kleinstiver, B.P., et al. 2015. Engineered CRISPR-Cas9 nucleases with altered PAM specificities. *Nature* 523: 481–485.
22. Mali, P., et al. 2013. CAS9 transcriptional activators for target specificity screening and paired nickases for cooperative genome engineering. *Nature Biotechnology* 31: 833–838.
23. Ran, F.A., et al. 2013. Double nicking by RNA-guided CRISPR Cas9 for enhanced genome editing specificity. *Cell* 154: 1380–1389.
24. Tsai, S.Q., et al. 2014. Dimeric CRISPR RNA-guided FokI nucleases for highly specific genome editing. *Nature Biotechnology* 32: 569–576.
25. Guilinger, J.P., D.B. Thompson, and D.R. Liu. 2014. Fusion of catalytically inactive Cas9 to FokI nuclease improves the specificity of genome modification. *Nature Biotechnology* 32: 577–582.
26. Wyvekens, N., V.V. Topkar, C. Khayter, J.K. Joung, and S.Q. Tsai. 2015. Dimeric CRISPR RNA-guided FokI-dCas9 nucleases directed by truncated gRNAs for highly specific genome editing. *Human Gene Therapy* 26: 425–431.
27. Deltcheva, E., et al. 2011. CRISPR RNA maturation by trans-encoded small RNA and host factor RNase III. *Nature* 471: 602–607.
28. Jiang, W., D. Bikard, D. Cox, F. Zhang, and L.A. Marraffini. 2013. RNA-guided editing of bacterial genomes using CRISPR-Cas systems. *Nature Biotechnology* 31: 233–239.
29. Sternberg, S.H., S. Redding, M. Jinek, E.C. Greene, and J.A. Doudna. 2014. DNA interrogation by the CRISPR RNA-guided endonuclease Cas9. *Nature* 507: 62–67.
30. Jinek, M. et al. 2013. RNA-programmed genome editing in human cells. *Elife* 2: e00471.
31. Slaymaker, I.M., et al. 2016. Rationally engineered Cas9 nucleases with improved specificity. *Science* 351: 84–88.
32. Fu, Y., et al. 2013. High-frequency off-target mutagenesis induced by CRISPR-Cas nucleases in human cells. *Nature Biotechnology* 31: 822–826.
33. Reyon, D., et al. 2012. FLASH assembly of TALENs for high-throughput genome editing. *Nature Biotechnology* 30: 460–465.
34. Knight, S.C., et al. 2015. Dynamics of CRISPR–Cas9 genome interrogation in living cells. *Science* 350: 823–826.
35. Lin, S., B.T. Staahl, R.K. Alla, and J.A. Doudna. 2014. Enhanced homology-directed human genome engineering by controlled timing of CRISPR/Cas9 delivery. *Elife* 3: e04766.
36. Yang, L., et al. 2013. Optimization of scarless human stem cell genome editing. *Nucleic Acids Research* 41: 9049–9061.
37. Chen, F., S.M. Pruett-Miller, Y. Huang, M. Gjoka, K. Duda, J. Taunton, T.N. Collingwood, M. Frodin, and G.D. Davis. 2011. High-frequency genome editing using ssDNA oligonucleotides with zinc-finger nucleases. *Nature Methods* 8: 753–755.
38. Cho, S.W., S. Kim, J.M. Kim, and J.S. Kim. 2013. Targeted genome engineering in human cells with the Cas9 RNA-guided endonuclease. *Nature Biotechnology* 31: 230–232.
39. Cho, S.W., J. Lee, D. Carroll, and J.S. Kim. 2013. Heritable gene knockout in Caenorhabditis elegans by direct injection of Cas9-sgRNA ribonucleoproteins. *Genetics* 195: 1177–1180.
40. Wang, H., et al. 2013. One-step generation of mice carrying mutations in multiple genes by CRISPR/Cas-mediated genome engineering. *Cell* 153: 910–918.
41. Sung, Y.H., et al. 2014. Highly efficient gene knockout in mice and zebrafish with RNA-guided endonucleases. *Genome Research* 24: 125–131.
42. Long, C., et al. 2016. Postnatal genome editing partially restores dystrophin expression in a mouse model of muscular dystrophy. *Science* 351: 400–403.
43. Swiech, L., et al. 2015. *Nature Biotechnology* 33: 102–106.
44. Ran, F.A., et al. 2015. In vivo genome editing using Staphylococcus aureus Cas9. *Nature* 520: 186–191.
45. Hur, J.K., et al. 2016. Targeted mutagenesis in mice by electroporation of Cpf1 ribonucleoproteins. *Nature Biotechnology* 34: 807–808.
46. Kleinstiver, B.P., et al. 2016. Genome-wide specificities of CRISPR-Cas Cpf1 nucleases in human cells. *Nature Biotechnology* 34: 869–874.

47. Komor, A.C., Y.B. Kim, M.S. Packer, J.A. Zuris, and D.R. Liu. 2016. Programmable editing of a target base in genomic DNA without double-stranded DNA cleavage. *Nature* 533: 420–424.
48. Gaudelli, N.M., et al. 2017. Programmable base editing of A*T to G*C in genomic DNA without DNA cleavage. *Nature* 551: 464–471.
49. Findlay, G.M., E.A. Boyle, R.J. Hause, J.C. Klein, and J. Shendure. 2014. Saturation editing of genomic regions by multiplex homology-directed repair. *Nature* 513: 120–123.
50. Kleinstiver, B.P., et al. 2015. Broadening the targeting range of Staphylococcus aureus CRISPR-Cas9 by modifying PAM recognition. *Nature Biotechnology* 33: 1293–1298.
51. Kim, Y.B., et al. 2017. Increasing the genome-targeting scope and precision of base editing with engineered Cas9-cytidine deaminase fusions. *Nature Biotechnology* 35: 371–376.
52. Gao, L., et al. 2017. Engineered Cpf1 variants with altered PAM specificities. *Nature Biotechnology* 35: 789–792.
53. Wu, D., et al. 2017. A label-free colorimetric isothermal cascade amplification for the detection of disease-related nucleic acids based on double-hairpin molecular beacon. *Analytica Chimica Acta* 957: 55–62.
54. Ermini, M.L., S. Mariani, S. Scarano, and M. Minunni. 2013. Direct detection of genomic DNA by surface plasmon resonance imaging: An optimized approach. *Biosensors & Bioelectronics* 40: 193–199.
55. Bao, Y.P., et al. 2005. SNP identification in unamplified human genomic DNA with gold nanoparticle probes. *Nucleic Acids Research* 33: e15–e15.
56. Gootenberg, J.S., et al. 2017. Nucleic acid detection with CRISPR-Cas13a/C2c2. *Science* 356: 438–442.
57. Li, S.-Y., et al. 2018. CRISPR–Cas12a-assisted nucleic acid detection. *Cell Discovery* 4: 20.
58. Pardee, K., et al. 2016. Rapid, low-cost detection of Zika virus using programmable biomolecular components. *Cell* 165: 1255–1266.
59. Chen, J.S., et al. 2018. CRISPR–Cas12a target binding unleashes indiscriminate single-stranded DNase activity. *Science* 360: 436–439.
60. Mekler, V., L. Minakhin, and K. Severinov. 2017. Mechanism of duplex DNA destabilization by RNA-guided Cas9 nuclease during target interrogation. *Proceedings of the National academy of Sciences of the United States of America* 114: 5443–5448.
61. Storhoff, J.J., et al. 2004. Gold nanoparticle-based detection of genomic DNA targets on microarrays using a novel optical detection system. *Biosensors & Bioelectronics* 19: 875–883.
62. Jung, Y.L., C. Jung, J.H. Park, M.I. Kim, and H.G. Park. 2013. Direct detection of unamplified genomic DNA based on photo-induced silver ion reduction by DNA molecules. *Chemical Communications* 49: 2350–2352.
63. Lau, H.Y., et al. 2017. Specific and sensitive isothermal electrochemical biosensor for plant pathogen DNA detection with colloidal gold nanoparticles as probes. *Science and Reports* 7: 38896.
64. Lee, I., et al. 2018. Simultaneous profiling of chromatin accessibility and methylation on human cell lines with nanopore sequencing. *Preprint at bioRxiv.* https://doi.org/10.1101/504993.
65. Martin, M., et al. 2016. WhatsHap: Fast and accurate read-based phasing. *Preprint at bioRxiv.* https://doi.org/10.1101/085050.
66. Li, H. 2011. A statistical framework for SNP calling, mutation discovery, association mapping and population genetical parameter estimation from sequencing data. *Bioinformatics* 27: 2987–2993.
67. Luo, R., et al. 2019. Clair: Exploring the limit of using a deep neural network on pileup data for germline variant calling. *Preprint at bioRxiv.* https://doi.org/10.1101/865782.
68. Simpson, J.T., et al. 2017. Detecting DNA cytosine methylation using nanopore sequencing. *Nature Methods* 14: 407–410.
69. Cebrian-Serrano, A., and B. Davies. 2017. CRISPR–Cas orthologues and variants: Optimizing the repertoire, specificity and delivery of genome engineering tools. *Mammalian Genome* 28: 247–261.

70. Hu, J.H., et al. 2018. Evolved Cas9 variants with broad PAM compatibility and high DNA specificity. *Nature* 556: 57–63.
71. Nishimasu, H., et al. 2018. Engineered CRISPR–Cas9 nuclease with expanded targeting space. *Science* 361: 1259–1262.
72. Rees, H.A., and D.R. Liu. 2018. Base editing: Precision chemistry on the genome and transcriptome of living cells. *Nature Reviews Genetics* 19: 770–778.
73. Nishida, K. et al. 2016. Targeted nucleotide editing using hybrid prokaryotic and vertebrate adaptive immune systems. *Science* 353: aaf8729–aaf8729.
74. Ma, Y., et al. 2016. Targeted AID-mediated mutagenesis (TAM) enables efficient genomic diversification in mammalian cells. *Nature Methods* 13: 1029–1035.
75. Hess, G.T., et al. 2016. Directed evolution using dCas9-targeted somatic hypermutation in mammalian cells. *Nature Methods* 13: 1036–1042.
76. Wang, X., et al. 2018. Efficient base editing in methylated regions with a human APOBEC3A–Cas9 fusion. *Nature Biotechnology* 36: 946–949.
77. Coelho, M.A., et al. 2018. BE-FLARE: A fluorescent reporter of base editing activity reveals editing characteristics of APOBEC3A and APOBEC3B. *BMC Biology* 16: 150.
78. St Martin, A. et al. 2018. A fluorescent reporter for quantification and enrichment of DNA editing by APOBEC–Cas9 or cleavage by Cas9 in living cells. *Nucleic Acids Research* 46: e84.
79. Martin, A.S., et al. 2019. A panel of eGFP reporters for single base editing by APOBEC-Cas9 editosome complexes. *Science and Reports* 9: 497.
80. Liu, Z. et al. 2019. Highly precise base editing with CC context-specificity using engineered human APOBEC3G-nCas9 fusions. bioRxiv https://www.biorxiv.org/content/https://doi.org/10.1101/658351v1.
81. Thuronyi, B.W.K., et al. 2019. Continuous evolution of base editors with expanded target compatibility and improved activity. *Nature Biotechnology* 37: 1070–1079.
82. Kim, Y.B., et al. 2017. Increasing the genome-targeting scope and precision of base editing with engineered Cas9–cytidine deaminase fusions. *Nature Biotechnology* 35: 371–376.
83. Grunewald, J., et al. 2019. Transcriptome-wide off-target RNA editing induced by CRISPR-guided DNA base editors. *Nature* 569: 433–437.
84. Gehrke, J.M., et al. 2018. An APOBEC3A–Cas9 base editor with minimized bystander and off-target activities. *Nature Biotechnology* 36: 977–982.
85. Zuo, E.S., et al. 2019. Cytosine base editor generates substantial off-target single-nucleotide variants in mouse embryos. *Science* 364: 289–292.
86. Jin, S.Z., et al. 2019. Cytosine, but not adenine, base editors induce genome-wide off-target mutations in rice. *Science* 364: 292–295.
87. Hazen, J.L., et al. 2016. The complete genome sequences, unique mutational spectra, and developmental potency of adult neurons revealed by cloning. *Neuron* 89: 1223–1236.
88. Milholland, B., et al. 2017. Differences between germline and somatic mutation rates in humans and mice. *Nature Communications* 8: 15183.
89. Dong, X., et al. 2017. Accurate identification of single-nucleotide variants in whole-genome-amplified single cells. *Nature Methods* 14: 491–493.
90. Lynch, M. 2010. Evolution of the mutation rate. *Trends in Genetics* 26: 345–352.
91. Rahbari, R., et al. 2016. Timing, rates and spectra of human germline mutation. *Nature Genetics* 48: 126–133.
92. Komor, A.C. et al. 2017. Improved base excision repair inhibition and bacteriophage Mu Gam protein yields C:G-to-T:A base editors with higher efficiency and product purity. *Science Advances* 3: eaao4774.
93. Rees, H.A., et al. 2017. Improving the DNA specificity and applicability of base editing through protein engineering and protein delivery. *Nature Communications* 8: 15790.
94. Kim, J., et al. 2006. Structural and kinetic characterization of Escherichia coli TadA, the wobble-specific tRNA deaminase. *Biochemistry* 45: 6407–6416.
95. Paquet, D., et al. 2016. Efficient introduction of specific homozygous and heterozygous mutations using CRISPR/Cas9. *Nature* 533: 125–129.

96. Kim, D., et al. 2017. Genome-wide target specificities of CRISPR RNA-guided programmable deaminases. *Nature Biotechnology* 35: 475–480.

97. Ryu, S.M., et al. 2017. Adenine base editing in mouse embryos and an adult mouse model of Duchenne muscular dystrophy. *Nature Biotechnology* 36: 536–539.

98. Paulk, N.K., et al. 2010. Adeno-associated virus gene repair corrects a mouse model of hereditary tyrosinemia in vivo. *Hepatology* 51: 1200–1208.

99. Aponte, J.L., et al. 2001. Point mutations in the murine fumarylacetoacetate hydrolase gene: Animal models for the human genetic disorder hereditary tyrosinemia type 1. *Proceedings of the National academy of Sciences of the United States of America* 98: 641–645.

100. Yin, H., et al. 2016. Therapeutic genome editing by combined viral and non-viral delivery of CRISPR system components in vivo. *Nature Biotechnology* 34: 328–333.

101. Song, C.Q., and W. Xue. 2018. CRISPR–Cas-related technologies in basic and translational liver research. *Nature Reviews. Gastroenterology & Hepatology* 15: 251–252.

102. Shao, Y., et al. 2018. Cas9-nickase-mediated genome editing corrects hereditary tyrosinemia in rats. *Journal of Biological Chemistry* 293: 6883–6892.

103. Wang, D., et al. 2018. Cas9-mediated allelic exchange repairs compound heterozygous recessive mutations in mice. *Nature Biotechnology* 36: 839–842.

104. Komor, A.C. et al. 2017. Improved base excision repair inhibition and bacteriophage Mu Gam protein yields C:G-to-T:A base editors with higher efficiency and product purity. *Science Advances* 3, eaao4774.

105. Yamanaka, S., K.S. Poksay, M.E. Balestra, G.Q. Zeng, and T.L. Innerarity. 1994. Cloning and mutagenesis of the rabbit ApoB mRNA editing protein. A zinc motif is essential for catalytic activity, and noncatalytic auxiliary factor(s) of the editing complex are widely distributed. *Journal of Biological Chemistry* 269: 21725–21734.

106. Navaratnam, N., et al. 1995. Evolutionary origins of apoB mRNA editing: catalysis by a cytidine deaminase that has acquired a novel RNA-binding motif at its active site. *Cell* 81: 187–195.

107. Teng, B.B. et al. 1999. Mutational analysis of apolipoprotein B mRNA editing enzyme (APOBEC1). Structure–function relationships of RNA editing and dimerization. *Journal of Lipid Research.* 40: 623–635.

108. Chen, Z., et al. 2010. Hypermutation induced by APOBEC-1 overexpression can be eliminated. *RNA* 16: 1040–1052.

109. MacGinnitie, A.J., S. Anant, and N.O. Davidson. 1995. Mutagenesis of apobec-1, the catalytic subunit of the mammalian apolipoprotein B mRNA editing enzyme, reveals distinct domains that mediate cytosine nucleoside deaminase, RNA binding, and RNA editing activity. *Journal of Biological Chemistry* 270: 14768–14775.

110. Gaudelli, N.M., et al. 2017. Programmable base editing of A•T to G•C in genomic DNA without DNA cleavage. *Nature* 551: 464–471.

111. Gao, X., et al. 2018. Treatment of autosomal dominant hearing loss by in vivo delivery of genome editing agents. *Nature* 553: 217–221.

112. Liu, Y., H.-I. Kao, and R.A. Bambara. 2004. Flap endonuclease 1: A central component of DNA metabolism. *Annual Review of Biochemistry* 73: 589–615.

113. Shen, M.W., et al. 2018. Predictable and precise template-free CRISPR editing of pathogenic variants. *Nature* 563: 646–651.

114. Koblan, L.W., et al. 2018. Improving cytidine and adenine base editors by expression optimization and ancestral reconstruction. *Nature Biotechnology* 36: 843–846.

115. Kleinstiver, B.P., et al. 2016. High-fidelity CRISPR-Cas9 nucleases with no detectable genome-wide off-target effects. *Nature* 529: 490–495.

116. Grünewald, J., et al. 2019. Transcriptome-wide off-target RNA editing induced by CRISPR-guided DNA base editors. *Nature* 569: 433–437.

117. Doman, J.L., A. Raguram, G.A. Newby, and D.R. Liu. 2020. Evaluation and minimization of Cas9-independent off-target DNA editing by cytosine base editors. *Nature Biotechnology* 38: 620–628.

118. Doench, J.G., et al. 2014. Rational design of highly active sgRNAs for CRISPR-Cas9-mediated gene inactivation. *Nature Biotechnology* 32: 1262–1267.
119. Wang, X., et al. 2015. Unbiased detection of off-target cleavage by CRISPR–Cas9 and TALENs using integrase-defective lentiviral vectors. *Nature Biotechnology* 33: 175–178.
120. Gratz, S.J., F.P. Ukken, C.D. Rubinstein, G. Thiede, L.K. Donohue, A.M. Cummings, et al. 2014. Highly specific and efficient CRISPR/Cas9-catalyzed homology-directed repair in Drosophila. *Genetics* 196: 961–971.
121. Ren, X., Z. Yang, J. Xu, J. Sun, D. Mao, Y. Hu, et al. 2014. Enhanced specificity and efficiency of the CRISPR/Cas9 system with optimized sgRNA parameters in Drosophila. *Cell Reports* 9: 1151–1162.
122. Heigwer, F., G. Kerr, and M. Boutros. 2014. E-CRISP: Fast CRISPR target site identification. *Nature Methods* 11: 122–123.
123. MacPherson, C.R., and A. Scherf. 2015. Flexible guide-RNA design for CRISPR applications using Protospacer Workbench. *Nature Biotechnology* 33: 805–806.
124. Xu, H., et al. 2015. Sequence determinants of improved CRISPR sgRNA design. *Genome Research* 25: 1147–1157.
125. Farboud, B., and B.J. Meyer. 2015. Dramatic enhancement of genome editing by CRISPR/Cas9 through improved guide RNA design. *Genetics* 199: 959–971.
126. Doench, J.G., et al. 2016. *Nature Biotechnology* 34: 184–191.
127. Hsu, P.D., et al. 2013. DNA targeting specificity of RNA-guided Cas9 nucleases. *Nature Biotechnology* 31: 827–832.
128. van Overbeek, M., et al. 2016. DNA repair profiling reveals nonrandom outcomes at Cas9-mediated breaks. *Molecular Cell* 63: 633–646.
129. Taheri-Ghahfarokhi, A., et al. 2018. Decoding non-random mutational signatures at Cas9 targeted sites. *Nucleic Acids Research* 46: 8417–8434.
130. Chakrabarti, A.M., et al. 2018. Target-specific precision of CRISPR-mediated genome editing. *Preprint at bioRxiv.* https://doi.org/10.1101/387027.
131. Bae, S., J. Kweon, H.S. Kim, and J.-S. Kim. 2014. Microhomology-based choice of Cas9 nuclease target sites. *Nature Methods* 11: 705–706.
132. Cho, S.W., et al. 2014. Analysis of off-target effects of CRISPR/Cas-derived RNA-guided endonucleases and nickases. *Genome Research* 24: 132–141.
133. Truong, L.N., et al. 2013. Microhomology-mediated end joining and homologous recombination share the initial end resection step to repair DNA double-strand breaks in mammalian cells. *Proceedings of the National academy of Sciences of the United States of America* 110: 7720–7725.
134. Jones, E., T. Oliphant, and P. Peterson. SciPy: Open source scientific tools for Python. SciPy http://www.scipy.org (2001, accessed 10 January 2018).
135. Shmakov, S., et al. 2015. Discovery and functional characterization of diverse class 2 CRISPR-Cas systems. *Molecular Cell* 60: 385–397.
136. Burstein, D., et al. 2017. New CRISPR–Cas systems from uncultivated microbes. *Nature* 542: 237–241.
137. Yan, W.X., et al. 2019. Functionally diverse type V CRISPR-Cas systems. *Science* 363: 88–91.
138. Hou, Z., et al. 2013. Efficient genome engineering in human pluripotent stem cells using Cas9 from Neisseria meningitidis. *Proceedings of the National academy of Sciences of the United States of America* 110: 15644–15649.
139. Kim, E., et al. 2017. In vivo genome editing with a small Cas9 orthologue derived from Campylobacter jejuni. *Nature Communications* 8: 14500.
140. Chatterjee, P., N. Jakimo, and J.M. Jacobson. 2018. Minimal PAM specificity of a highly similar SpCas9 ortholog. *Science Advances* 4: eaau0766.
141. Zetsche, B., et al. 2015. Cpf1 is a single RNA-guided endonuclease of a class 2 CRISPR-Cas system. *Cell* 163: 759–771.
142. Abudayyeh, O.O., et al. 2017. RNA targeting with CRISPR–Cas13. *Nature* 550: 280–284.
143. Konermann, S., et al. 2018. Transcriptome engineering with RNA-targeting type VI-D CRISPR effectors. *Cell* 173: 665-676.e14.

144. Zalatan, J.G., et al. 2015. Engineering complex synthetic transcriptional programs with CRISPR RNA scaffolds. *Cell* 160: 339–350.
145. Sternberg, S.H., B. LaFrance, M. Kaplan, and J.A. Doudna. 2015. Conformational control of DNA target cleavage by CRISPR–Cas9. *Nature* 527: 110–113.

Chapter 4
Application of CRISPR-Based Technology in Medical Research and Disease Treatment

Ziheng Zhang, Ping Wang, and Ji-Long Liu

Introduction

Seeking treatments for human diseases is an important goal and motivation of biomedical research. Over thousands of years, humans have gradually formed a series of treatment plans for diseases, which have gradually evolved into modern medicine. Although many common diseases have been overcome, a cure for diseases caused by genetic mutations, including congenital genetic diseases, is still lacking. To overcome genetic diseases caused by DNA mutations, new strategies must be developed. An example that has received much attention is gene therapy, which correct or compensate diseases caused by defects and abnormal genes by introducing foreign normal genes into target cells to achieve the purpose of treatment. Although these methods have been proven effective and are undergoing clinical trials, the ideal treatment is to fundamentally correct the mutated DNA.

The powerful gene editing capabilities of CRISPR have enabled us to focus on the application of CRISPR in disease treatment and medical research. For diseases caused by genetic mutations carried by congenital abnormalities, the emergence of CRISPR provides an unprecedented therapeutic strategy with great potential. In particular, the development of base editor, which can achieve precise editing of a single specific base without causing DNA double-strand breaks, provides a great strategy for correcting a series of congenital genetic diseases caused by genetic mutations. In addition, CRISPR has also been widely used in research in the medical field, including the construction of a series of animal models through gene knockout or mutation. The gene-based loss of function or gain of function CRISPR screening

Z. Zhang
School of Life Science and Technology, ShanghaiTech University, Shanghai, China

P. Wang
University Library, ShanghaiTech University, Shanghai, China

J.-L. Liu (✉)
School of Life Science and Technology, ShanghaiTech University, Shanghai, China

has been widely used in recent years, which directly leads to the acquisition of many potential drug targets that have not been reached by previous technologies. In this chapter, we introduce some applications of CRISPR in medical research, including the discovery of potential drug targets by CRISPR-based screening, and the use of CRISPR in the treatment of genetic diseases.

Machine generated keywords: tumour, cancer, antigen, signal, lipid, organoid, factor, cancer cell, screen, receptor, vivo, response, pathway, patient, droplet.

4.1 CRISPR in Biomedical Research

Machine generated keywords: tumour, cancer, antigen, signal, organoid, factor, cancer cell, lipid, screen, receptor, response, pathway, cell line, vivo, efficacy.

Modeling colorectal cancer using CRISPR-Cas9–mediated engineering of human intestinal organoids.

https://doi.org/10.1038/nm.3802

Main

Because these niche factors influence most if not all of the pathways that are mutated in CRC, driver pathway mutations may modulate carcinogenesis by conferring selective growth advantages to ISCs [1].

Here, to examine the influence of various driver pathway mutations on human colorectal carcinogenesis, we prospectively introduced multiple driver pathway mutations into human normal or adenoma-derived intestinal organoids.

We established an efficient system to introduce multiple driver gene mutations into the human colonic epithelium by niche factor–modulated culture conditions, enabling clonal expansion of mutant stem cells without antibiotic selection.

Most of CRC (>90%) as well as genetically engineered mouse CRC models possess fewer than five driver pathway alterations, suggesting that the non-mutated niche signaling in CRC could be complemented by adjacent non-epithelial mesenchymal or immune cells.

Methods

For the organoid formation assay and before genome editing, organoids were disso-ciated into single cells with TrypLE Express (Life Technologies) and 1000 cells were cultured in a 48-well plate under the above culture conditions for 10 d. To prevent anoikis, 10 μM Y-27632 was included in the culture medium for the first 2 d (ref 2).

After dissociating the organoids with TrypLE express, cells were electroporated with 10 μg of CRISPR-Cas9 plasmid (Neppagene).

Tumor organoids were expanded for gene expression analysis or copy number analysis.

For gene expression microarray analysis, 500 ng of whole RNA was extracted from organoids using the RNeasy Mini Plus kit (Qiagen).

The GFP-expressing organoids were isolated from Matrigel using Cell Recovery Solution (BD Biosciences) and mechanically dissociated into small clusters of cells.

We euthanized 2 out of 60 mice with implanted CRC organoids due to the tumor burden during the experimental period (2 months) and excluded these mice from the data analysis.

Acknowledgements

A machine generated summary based on the work of Matano, Mami; Date, Shoichi; Shimokawa, Mariko; Takano, Ai; Fujii, Masayuki; Ohta, Yuki; Watanabe, Toshiaki; Kanai, Takanori; Sato, Toshiro 2015 in Nature Medicine.

Sequential cancer mutations in cultured human intestinal stem cells.

https://doi.org/10.1038/nature14415

Abstract-Summary

Both mouse and human intestinal stem cells can be cultured in medium containing the stem-cell-niche factors WNT, R-spondin, epidermal growth factor (EGF) and noggin over long time periods as epithelial organoids that remain genetically and phenotypically stable.

We utilize CRISPR/Cas9 technology for targeted gene modification of four of the most commonly mutated colorectal cancer genes (APC, P53 (also known as TP53), KRAS and SMAD4) in cultured human intestinal stem cells.

Mutant organoids can be selected by removing individual growth factors from the culture medium.

Quadruple mutants grow independently of all stem-cell-niche factors and tolerate the presence of the P53 stabilizer nutlin-3.

Main

Progression is thought to occur through activating mutations in the EGF receptor (EGFR) pathway and inactivating mutations in the P53 and transforming growth factor (TGF)-β pathways [2].

Using mouse models, Lgr5[+]-intestinal stem cells were identified as cells of origin for intestinal neoplasia and were shown to fuel effective tumour growth [3–5].

A recent study has shown that deregulation (by retroviral expression of short hairpin RNAs (shRNAs) or cDNA) of APC, P53, KRAS and SMAD4 is sufficient for transformation of cultured mouse colon into tumours with adenocarcinoma-like histology [6].

Comparable human in vitro model systems to study tumour initiation and progression have not been developed.

Sequential introduction of CRC mutations

We set out to utilize CRISPR/Cas9-mediated genome editing [7–9] to introduce four of the most frequent CRC mutations in human small intestinal organoid stem cell cultures.

Since loss of APC is generally considered to be an early event in CRC [2], we first introduced inactivating mutations in APC.

To obtain clonal cultures, individual organoids were expanded.

We introduced inactivating mutations in P53 in APC knockout (APC^{KO}) intestinal organoids.

Quadruple mutants do not need niche factors

As with all sgRNAs, SMAD4 sgRNAs target the mutation hotspot region, encoding the MH2 domain required for SMAD4 activity [10, 11].

Although this analysis was limited, in combination with the analysis of multiple independent clonal organoids, the results indicated that the observed effects were not due to off-target effects.

Quadruple mutants grow as invasive carcinomas

We investigated whether our engineered organoids were tumorigenic in vivo.

[Section 5]

To verify that the observed CIN results in aneuploidy, we next counted chromosome numbers.

Loss of both APC and P53 had the most dramatic effect on CIN and aneuploidy.

We show that the combined loss of APC and P53 is sufficient for the appearance of extensive aneuploidy.

CRC mutations in human colon organoids

We believe that our model reflects the in vivo situation more closely than any other in vitro human CRC model so far.

Upon oncogenic mutation of KRAS, APC, P53 and SMAD4, human gut stem cell organoids can grow in the absence of all stem-cell-niche factors and in the presence of the P53 stabilizer nutlin-3 in vitro and as tumours with invasive carcinoma features in vivo.

Methods

For selection of KRAS^{G12D} mutants, organoids were grown in culture medium lacking EGF and containing 0.5–1.0 μM of gefitinib (Selleck Chemicals).

Organoid treatments: WNT/R-spondin withdrawal, 48 h; nutlin-3 10 μM, 24 h. Primer sequences: AXIN2_for, 5′-AGCTTACATGAGTAATGGGG-3′, AXIN2_rev, 5′-AATTCCATCTACACTGCTGTC-3′; P21_for, 5′-TACCCTTGTGCCTCGCTCAG-3′, P21_rev, 5′-GAGAAGATCAGCCGGCGTTT-3′; GAPDH_for, 5′-TGCACCACCAACTGCTTAGC-3′, GAPDH_rev, 5′-GGCATGGACTGTGGTCATGAG-3′. Samples were lysed using RIPA buffer (50 mM Tris–HCl pH 8.0, 150 mM NaCl, 0.1% SDS, 0.5% Na-Deoxycholate, 1% NP-40) containing Complete protease inhibitors (Roche).

After trypsinization, 200,000 cells were resuspended in 50 μl of medium containing 2 × required growth factors, mixed with Matrigel (BD Biosciences) at a 1:1 ratio and injected subcutaneously into NOD scid gamma (NSG; NOD.Cg-Prkdc^{scid} Il2rg^{tm1Wjl}/SzJ) mice ($\geq$6 injections per organoid line).

For karyotyping, organoids were treated with 0.1 µg ml^{-1} colcemid (Gibco) for 16 h. Cultures were washed and dissociated into single cells using TrypLE (Gibco) and processed as described [12].

Acknowledgements

A machine generated summary based on the work of Drost, Jarno; van Jaarsveld, Richard H.; Ponsioen, Bas; Zimberlin, Cheryl; van Boxtel, Ruben; Buijs, Arjan; Sachs, Norman; Overmeer, René M.; Offerhaus, G. Johan; Begthel, Harry; Korving, Jeroen; van de Wetering, Marc; Schwank, Gerald; Logtenberg, Meike; Cuppen, Edwin; Snippert, Hugo J.; Medema, Jan Paul; Kops, Geert J. P. L.; Clevers, Hans 2015 in Nature.

In vivo *genome editing* via *CRISPR/Cas9 mediated homology-independent targeted integration.*

https://doi.org/10.1038/nature20565

Main

Regardless, however, NHEJ-mediated targeted transgene integration in post-mitotic cells has yet to be determined, especially in vivo in adult tissues such as the brain.

We aim to develop a robust NHEJ-based homology-independent strategy for targeted integration of transgenes in both dividing and non-dividing cells.

We tested HITI in non-dividing cells in vitro.

All HITI vectors allowed efficient GFP knock-in in non-dividing primary neurons.

To demonstrate the efficacy of HITI in post-mitotic cells in vivo, we generated inducible Tubb3–GFP HITI targeting constructs where Cre-dependent Cas9 expression is under the control of tamoxifen (TAM).

These results demonstrate the utility of HITI for transgene knock-in in a variety of somatic tissues, including non-dividing cells, in vivo.

The ability to use HITI for in vivo targeted transgene insertion into post-mitotic neurons is unprecedented and will help advance basic and translational neuroscience research.

Methods

Two days after electroporation, the cells were dissociated by TrypLE, and Cas9 and gRNA expressing cells were sorted out as GFP/mCherry double-positive cells by BD influx cell sorter (BD), and genomic DNA extracted with DNeasy Blood & Tissue kit.

Free-floating sections were incubated at 4 °C for 16–48 h with goat anti-GFP (Rockland) primary antibodies in PBS/0.5% normal donkey serum/0.1% Triton X-100, followed by the appropriate secondary antibodies conjugated with Alexa Fluor 488 at room temperature for 2–3 h. Sections were counterstained with 10 µM DAPI in PBS for 30 min to visualize cell nuclei.

1 μl of DNA mixture, containing the pCAG-1BPNLS-Cas9-1BPNLS (0.5 μg μl^{-1}), mouse Tubb3 gene target pCAGmCherry-gRNA (0.5 μg μl^{-1}) and either donor cut-only control donor (Tubb3-MC-scramble), minicircle donor (Tubb3-MC), 2-cut (Tubb3-2c) or HDR donor (Tubb3-HDR) vectors (0.8 μg μl^{-1}) was injected into the hemisphere of the fetal brain.

Acknowledgements

A machine generated summary based on the work of Suzuki, Keiichiro; Tsunekawa, Yuji; Hernandez-Benitez, Reyna; Wu, Jun; Zhu, Jie; Kim, Euiseok J.; Hatanaka, Fumiyuki; Yamamoto, Mako; Araoka, Toshikazu; Li, Zhe; Kurita, Masakazu; Hishida, Tomoaki; Li, Mo; Aizawa, Emi; Guo, Shicheng; Chen, Song; Goebl, April; Soligalla, Rupa Devi; Qu, Jing; Jiang, Tingshuai; Fu, Xin; Jafari, Maryam; Esteban, Concepcion Rodriguez; Berggren, W. Travis; Lajara, Jeronimo; Nuñez-Delicado, Estrella; Guillen, Pedro; Campistol, Josep M.; Matsuzaki, Fumio; Liu, Guang-Hui; Magistretti, Pierre; Zhang, Kun; Callaway, Edward M.; Zhang, Kang; Belmonte, Juan Carlos Izpisua 2016 in Nature.

Targeting a CAR to the TRAC locus with CRISPR/Cas9 enhances tumour rejection.

https://doi.org/10.1038/nature21405

Main

These in vivo findings thus not only demonstrated the superior anti-tumour activity of TRAC-CAR T cells, but also forged a link between tumour control, T-cell differentiation and exhaustion, and CAR expression levels.

These results indicated that the improved efficacy of TRAC-CAR T cells is related to its CAR expression level by reducing tonic signalling and delaying T-cell differentiation upon stimulation.

Although TRAC-LTR directed lower baseline CAR expression than RV-CAR and averted the tonic signalling, the LTR still promoted from within the TRAC locus the same differentiation pattern as RV-CAR.

To T cells with higher CAR expression, the TRAC-CAR profile correlated with decreased T-cell differentiation and exhaustion, resulting in superior tumour eradication.

Our kinetic measurements of antigen-induced CAR internalization and degradation revealed differential recovery of cell-surface CAR depending on the enhancer/promoter elements driving CAR expression.

Methods

We used the same protocol and obtained similar cytotoxicity and specificity of both methods and the CAR T cells obtained were not discernable in term of activity and proliferation (data not shown).

Based on a pAAV-GFP backbone (Cell Biolabs) we designed and cloned the pAAV-TRAC-1928z containing 1.9 kb of genomic TRAC (amplified by PCR) flanking the gRNA targeting sequences, a self-cleaving P2A peptide in frame with the first exon of TRAC followed by the 1928z CAR used in clinical trials [13].

NALM-6 cells were transduced to express firefly luciferase-GFP and NIH/3T3 cells transduced to express human CD19 (refs 14, 15).

The effector (E) and tumour target (T) cells were co-cultured in triplicates at the indicated E/T ratio using black-walled 96- well plates with 1×10^5 target cells in a total volume of 100 µl per well in NALM-6 Medium.

Acknowledgements

A machine generated summary based on the work of Eyquem, Justin; Mansilla-Soto, Jorge; Giavridis, Theodoros; van der Stegen, Sjoukje J. C.; Hamieh, Mohamad; Cunanan, Kristen M.; Odak, Ashlesha; Gönen, Mithat; Sadelain, Michel 2017 in Nature.

CRISPR/Cas9-mediated PD-1 disruption enhances anti-tumor efficacy of human chimeric antigen receptor T cells.

https://doi.org/10.1038/s41598-017-00462-8

Abstract-Summary

We show that programmed death ligand 1 (PD-L1) expression on tumor cells can render human CAR T cells (anti-CD19 4-1BBζ) hypo-functional, resulting in impaired tumor clearance in a sub-cutaneous xenograft model.

To overcome this suppressed anti-tumor response, we developed a protocol for combined Cas9 ribonucleoprotein (Cas9 RNP)-mediated gene editing and lentiviral transduction to generate PD-1 deficient anti-CD19 CAR T cells.

Pdcd1 (PD-1) disruption augmented CAR T cell mediated killing of tumor cells in vitro and enhanced clearance of PD-L1+ tumor xenografts in vivo.

Extended:

Combined, our results indicate that CRISPR-mediated targeted disruption of the Pdcd1 locus can enhance in vivo anti-tumor efficacy of human CAR T cells.

Introduction

We first explored whether expression of the inhibitory receptor PD-L1 on tumor cells could suppress highly potent anti-CD19 CAR T cells.

We subsequently developed a protocol for combined Cas9-based gene disruption and lentiviral transduction to generate gene modified human CAR T cells, and then tested whether Cas9-mediated disruption of PD-1 in CAR T cells improved anti-tumor efficacy in vitro and in vivo using a xenograft tumor model.

Despite the clear role of PD-1/PD-L1 in regulating endogenous anti-tumor responses, the impact of inhibitory receptors on CAR T cell function remains largely unexplored.

We tested whether Cas9 RNP mediated disruption of the endogenous Pdcd1 locus in primary human CAR T cells enhances anti-tumor efficacy.

We found that PD-L1 expression on tumor cells impaired CAR T cell mediated killing in vitro and tumor clearance in vivo in a xenograft model.

Results

We utilized a second generation anti-CD19 4-1BBζ CAR (aCD19 CAR) for these studies because it has displayed potent anti-tumor activity in clinical studies [16] and we hypothesized that it would present a stringent challenge for PD-L1 mediated suppression of CAR T cell function.

Combined, these results suggest that PD-L1 expression on tumor cells can directly impair the function and lytic capacity of primary human CD8+ T cells expressing the anti-CD19 4-1BBζ CAR.

We also tested the cytolytic capacity of PD-1 edited anti-CD19 CAR T cells against both CD19+ and CD19+ PD-L1+ targets.

These results suggest that Pdcd1 disruption can rescue functional defects induced in CD8+ anti-CD19 4-1BBζ CAR T cells by PD-L1+ tumor cells.

NSG mice were injected subcutaneously with CD19+ PD-L1+ K562 cells and tumors were established before injection with either PD-1 edited or Cas9 control (non-edited) anti-CD19 CAR T cells (4×10^6 CD4+ CAR+ and 4×10^6 CD8+ CAR+ cells).

Discussion

We show that tumor-specific PD-L1 expression can render second generation anti-CD19 4-1BBζ CAR T cells hypo-functional in vitro and impair tumor clearance in vivo in a subcutaneous tumor xenograft model.

These results demonstrate the inhibitory role of the PD-1/PD-L1 axis on CAR T cell anti-tumor function, and demonstrate proof-of-principle for Cas9-based gene editing to enhance CAR T cell efficacy.

These results suggest it is possible that robust deletion of Pdcd1 could enhance short-term functions of CAR T cells but ultimately render edited cells more susceptible to exhaustion/impaired function (for example, if the tumor is not rapidly cleared).

Further characterization of T cells in the CD19+ vs. CD19+ PD-L1+ model could help unravel the mechanisms by which tumor-specific PD-L1 expression impairs CAR T cell function.

The enhanced anti-tumor functionality of Pdcd1 edited CAR T cells might allow physicians to administer fewer cells to achieve equivalent therapeutic effects.

Materials and Methods

To generate the CD19+ PD-L1+ cell line, 1×10^5 CD19+ K562 cells were transduced with lentiviral supernatant for 24 h.

Rested CD8+ anti-CD19 CAR T cells were incubated with CD19+ or CD19+ PD-L1+ K562s at an effector:target (E:T) ratio of 1:2 in T cell medium containing anti-CD107a, with 1×10^5 CAR+ T cells seeded per well of 96-well plate.

CD19+ or CD19+ PD-L1+ (Ag^+) targets were mixed 1:1 with Ag^- targets and 1×10^5 total K562 cells seeded per well in 96-well plates.

For restimulation experiments, rested CD8+ CAR T cells were re-stimulated 1:1 with CD19+ K562 cells for 48 h prior to staining for CD69 and PD-1, followed by flow cytometry analysis.

When tumors reached 100–1000 mm^3 in volume, 2–7.2 × 10^6 CAR+ CD4+ and 2–4.1 × 10^6 CAR+ CD8+ control or PD-1 edited CAR T cells were injected intravenously as indicated in various figures.

Acknowledgements

A machine generated summary based on the work of Rupp, Levi J.; Schumann, Kathrin; Roybal, Kole T.; Gate, Rachel E.; Ye, Chun J.; Lim, Wendell A.; Marson, Alexander 2017 in Scientific Reports.

In vivo *CRISPR screening identifies Ptpn2 as a cancer immunotherapy target.*

https://doi.org/10.1038/nature23270

Abstract-Summary

We use a pooled in vivo genetic screening approach using CRISPR–Cas9 genome editing in transplantable tumours in mice treated with immunotherapy to discover previously undescribed immunotherapy targets.

Tumours were sensitized to immunotherapy by deletion of genes involved in several diverse pathways, including NF-κB signalling, antigen presentation and the unfolded protein response.

Deletion of the protein tyrosine phosphatase PTPN2 in tumour cells increased the efficacy of immunotherapy by enhancing interferon-γ-mediated effects on antigen presentation and growth suppression.

Genetic screens in tumour models can identify new immunotherapy targets in unanticipated pathways.

Main

Although cancer cells could, in theory, express many more genes that regulate their response or resistance to tumour immunity, strategies to systematically discover such genes are lacking.

These approaches include pooled genetic screens using CRISPR–Cas9-mediated genome editing that simultaneously test the role of a large number of genes on tumour cell growth, viability or drug resistance [17].

We use a pooled loss-of-function in vivo genetic screening approach that uses CRISPR–Cas9 genome editing to discover genes that increase sensitivity or cause resistance to immunotherapy in a mouse transplantable tumour model.

[Section 2]

Inspection of the list of genes targeted by sgRNAs that are depleted from tumours treated with immunotherapy revealed the known immune evasion molecule PD-L1, indicating that loss of PD-L1 increased the sensitivity of tumour cells to immune attack.

Defects in the IFNγ pathway induces resistance

We next sought to determine why IFNγ-pathway mutant tumours were resistant to immunotherapy.

Gene targets that increase efficacy of immunotherapy

These genes were Ptpn2, a phosphatase involved in multiple signalling processes; H2-T23, a non-classical MHC-I gene; Ripk1, a kinase that regulates cell death and inflammation; and Stub1, an E3 ubiquitin ligase involved in the regulation of the unfolded protein response.

This suggests that loss of function of these genes renders tumour cells more sensitive to immunotherapy, but does not alter their cell growth or survival in the absence of T cells.

These results indicate that Qa-1b functions as an immune evasion molecule in tumours and that loss of function of H2-T23 improves tumour immunity.

[Section 5]

To determine whether PTPN2 amplification was associated with immunotherapy resistance in human cancer, we examined exome-sequencing data from patients treated with blockade of PD-1, PD-L1 or CTLA-4.

[Section 6]

To test whether loss of Ptpn2 made tumour cells more recognizable to T cells, we cultured OVA-expressing Ptpn2-null or control cells with OT-I CD8$^+$ T cells that recognize the SIINFEKL epitope.

To test whether Ptpn2-null tumour cells were more sensitive to the presence of T cells, we co-cultured mixed populations of OVA-expressing Ptpn2-null or wild-type B16 cells with antigen-specific CD8$^+$ T cells for six days.

[Section 7]

We first assayed STAT1 activity in Ptpn2-null cells in response to IFNγ stimulation.

These results demonstrate that IFNγ alone is sufficient to cause a growth disadvantage in both mouse and human tumour cells that lack Ptpn2/PTPN2.

[Section 8]

The mechanism by which Ptpn2 deficiency sensitizes tumour cells to immunotherapy is dependent on the sensing of IFNγ.

Discussion

Loss of Ptpn2 in tumour cells increased IFNγ signalling and antigen presentation to T cells, and amplified growth arrest in response to cytokines, suggesting that its therapeutic inhibition may potentiate the effect of immunotherapies that invoke an IFNγ response.

Loss of function of Qa-1b (the non-classical MHC molecule encoded by H2-T23, or HLA-E in humans) increased sensitivity of tumour cells to immune attack.

Loss of function of Qa-1b may increase the immune response to tumours by de-repression of T- or NK-cell function [18, 19] or by limiting the stimulation of CD8$^+$ T regulatory cells [20], either of which would make blockade of HLA-E an attractive immunotherapeutic approach.

Tumours were also sensitized to immunotherapy by deletion of Ripk1, as well as by loss of function of members of the linear ubiquitin assembly complex (LUBAC), Birc2, Rbck1 or Rnf31, that responsible for ubiquitination and stabilization of the signalling complex that includes Ripk1.

Methods

Mice were vaccinated with 1.0×10^6 GM-CSF-secreting B16 (GVAX) cells (provided by G. Dranoff) that had been irradiated with 3500 Gy on days 1 and 4 to elicit an anti-tumour immune response.

For B16 cells, 5×10^5 cells were plated in a well of a six-well plate and were transfected the following day using 2 µg of pX459 plasmid DNA and Turbofect (3:1 ratio, Thermo Fisher Scientific).

For restimulation, 1×10^5 OVA-expressing Ptpn2-null or control sgRNA-transfected B16 tumour cells were first plated in 24-well plates and stimulated with recombinant mouse IFNγ overnight to induce ovalbumin surface expression.

1×10^6 pre-activated OT-I T cells were then added to the wells on the next day and co-cultured with the B16 tumour cells for 2–3 h. Subsequently, $1 \times$ brefeldin (eBiosciences) was added to the cultures for 4 h to inhibit intracellular protein transport.

Acknowledgements

A machine generated summary based on the work of Manguso, Robert T.; Pope, Hans W.; Zimmer, Margaret D.; Brown, Flavian D.; Yates, Kathleen B.; Miller, Brian C.; Collins, Natalie B.; Bi, Kevin; LaFleur, Martin W.; Juneja, Vikram R.; Weiss, Sarah A.; Lo, Jennifer; Fisher, David E.; Miao, Diana; Van Allen, Eliezer; Root, David E.; Sharpe, Arlene H.; Doench, John G.; Haining, W. Nicholas 2017 in Nature.

Computational correction of copy number effect improves specificity of CRISPR–Cas9 essentiality screens in cancer cells.

https://doi.org/10.1038/ng.3984

Abstract-Summary

CERES is a new computational method to estimate gene-dependency levels from CRISPR–Cas9 essentiality screens while accounting for copy number effects and variable sgRNA activity.

Main

We and others have recently observed that measurements of cell proliferation in genome-scale CRISPR–Cas9 loss-of-function screens are influenced by the genomic copy number of the region targeted by the single guide RNA (sgRNA)–Cas9 complex [21–23].

To quantify the extent to which this sgRNA-level effect translates into false-positive gene dependencies, we ranked the genes in each cell line by the average depletion of their targeting sgRNAs (average guide score).

To confirm this finding by using a complementary assay, we analyzed this set of genes in a data set derived from genome-scale RNA-interference screens across 501 cancer cell lines [24].

We introduce a large set of uniformly performed CRISPR–Cas9 essentiality screens of cancer cell lines, propose a methodology to estimate gene dependency while removing false positives due to copy number effects, and demonstrate the power of these two resources in identifying genetic vulnerabilities of cancer.

Method

The constants M, N, and K in the objective function are, respectively, the total number of sgRNAs, cell lines, and genes in the data set.

Given a collection of CRISPR screening data, let N be the number of sgRNAs, M be the number of cell lines, and K be the number of targeted genes in the data set.

For each cell line, average guide scores were regressed against gene-level copy number data by using a linear model.

For each number p in the set {1, 2, 4, 8, 16, 32, 64}, we ran 342/p trials (rounded up to the nearest integer), such that each cell line appeared once in each run of size p. For each p and each cell line, we evaluated the harmonic mean of precision and recall (referred to as the F1 measure) at the point of equiprobability between the essential and nonessential gene classes.

Acknowledgements

A machine generated summary based on the work of Meyers, Robin M; Bryan, Jordan G; McFarland, James M; Weir, Barbara A; Sizemore, Ann E; Xu, Han; Dharia, Neekesh V; Montgomery, Phillip G; Cowley, Glenn S; Pantel, Sasha; Goodale, Amy; Lee, Yenarae; Ali, Levi D; Jiang, Guozhi; Lubonja, Rakela; Harrington, William F; Strickland, Matthew; Wu, Ting; Hawes, Derek C; Zhivich, Victor A; Wyatt, Meghan R; Kalani, Zohra; Chang, Jaime J; Okamoto, Michael; Stegmaier, Kimberly; Golub, Todd R; Boehm, Jesse S; Vazquez, Francisca; Root, David E; Hahn, William C; Tsherniak, Aviad 2017 in Nature Genetics.

Promoter-bound METTL3 maintains myeloid leukaemia by m^6A-dependent translation control.

https://doi.org/10.1038/nature24678

Main

This suggests that METTL3 regulates the expression of mRNAs derived from its chromatin target genes.

To identify RNAs methylated by METTL3 in AML cells, we performed RNA-immunoprecipitation linked to high throughput sequencing using an m^{6}A-specific antibody (m^{6}A–IP).

To better characterize how METTL3 regulates translation, we focused on two genes encoding the transcription factors SP1 and SP2, which have promoters occupied by METTL3.

In this pathway, METTL3 is stably recruited by CEBPZ to promoters of a specific set of active genes, resulting in m^{6}A methylation of the respective mRNAs and increased translation.

The pathway described here is critical for AML leukaemia, as three of its components are required for AML cell growth: (i) the m^{6}A RNA methyltransferases METTL3; (ii) the transcription factor CEBPZ, which targets this enzyme to promoters; and (iii) SP1, whose translation is dependent upon the m^{6}A modification by METTL3.

Methods

RN2C cells or NIH3t3 mouse fibroblast cells were infected with LRG lentiviral vectors expressing GFP and a single gRNA targeting the catalytic domain of the indicated RNA enzymes and controls.

MOLM13 or THP1 cells (5×10^5) were infected as described above using PLKO-TETon-Puro lentiviral vectors expressing shRNAs against the coding sequence of human METTL3, CEBPZ or a scrambled control.

Total RNA was isolated from MOLM13 control or METTL3-KD cells (two independent biological replicates for each shRNA) eight days after doxycycline administration using the RNAeasy midi kit (Quiagen).

5×10^7 MOLM13 control (two pooled independent biological replicates) or METTL3-KD cells (two pooled independent biological replicates for each shRNA) were treated 5 or 8 days after doxycycline induction with 0.1 mg ml^{-1} of cycloheximide for 1 min and the RPF fraction of mRNA was isolated following the manufacturer's instruction.

Acknowledgements

A machine generated summary based on the work of Barbieri, Isaia; Tzelepis, Konstantinos; Pandolfini, Luca; Shi, Junwei; Millán-Zambrano, Gonzalo; Robson, Samuel C.; Aspris, Demetrios; Migliori, Valentina; Bannister, Andrew J.; Han, Namshik; De Braekeleer, Etienne; Ponstingl, Hannes; Hendrick, Alan; Vakoc, Christopher R.; Vassiliou, George S.; Kouzarides, Tony 2017 in Nature.

Selective silencing of euchromatic L1s revealed by genome-wide screens for L1 regulators.

https://doi.org/10.1038/nature25179

Main

A subset of regulators may function in a cell-type specific manner that is not captured by either K562 or HeLa screens; essential genes with strong negative effects on cell growth may have dropped out; or regulators that strictly require native L1 UTR sequences may have been missed owing to our reporter design.

Our combined screens identify many novel candidates for the control of L1 retrotransposition in human cells and provide a rich resource for mechanistic studies of transposable elements.

We further investigated three candidate transcriptional regulators of L1: MORC2, TASOR and MPP8.

Chromatin immunoprecipitation followed by sequencing (ChIP–seq) from K562 cells and hES cells demonstrated that MORC2, MPP8 and TASOR co-bind genomic regions characterized by specific L1 instances.

To understand the role of transcription in the HUSH and MORC2 targeting of L1s, we investigated MORC2 and MPP8 occupancy at the inducible L1 transgene.

Methods

The Cas9/L1-G418^R cells were lentivirally infected with a genome-wide sgRNA library as described previously [25], containing approximately 200,000 sgRNAs targeting 20,549 protein-coding genes and 13,500 negative-control sgRNAs at a multiplicity of infection of 0.3–0.4 (as measured by the mCherry fluorescence from the lentiviral vector), and selected for lentiviral integration using puromycin (1 μg ml^{-1}) for 3 days as the cultures were expanded for the screens.

After puromycin selection (1 μg ml^{-1} for 3 days) and expansion, 40 million (approximately 9000 coverage per library element) cells were dox-induced for 10 days in replicate, recovered for 1 day, and split for 7-day G418-selection and non-selection conditions, with a logarithmic growth (500,000 cells per ml) maintained as in the K562 genome-wide screen.

For the genome-wide screen, approximately 200 $\times$ 10^6 Cas9/L1-G418^R HeLa cells (around 1000 $\times$ coverage of the sgRNA library) were dox-induced for 10 days in replicate, recovered for 1 day, and split for 8-day G418-selection and non-selection conditions, with cells being split every other day to maintain the sgRNA library at a minimum of around 350 $\times$ coverage.

Acknowledgements

A machine generated summary based on the work of Liu, Nian; Lee, Cameron H.; Swigut, Tomek; Grow, Edward; Gu, Bo; Bassik, Michael C.; Wysocka, Joanna 2017 in Nature.

A multiprotein supercomplex controlling oncogenic signalling in lymphoma.

https://doi.org/10.1038/s41586-018-0290-0

Abstract-Summary

B cell receptor (BCR) signalling has emerged as a therapeutic target in B cell lymphomas, but inhibiting this pathway in diffuse large B cell lymphoma (DLBCL) has benefited only a subset of patients [26].

Autoantigens drive BCR-dependent activation of NF-κB in ABC DLBCL through a kinase signalling cascade of SYK, BTK and PKCβ to promote the assembly of the CARD11–BCL10–MALT1 adaptor complex, which recruits and activates IκB kinase [27–29].

Genome sequencing revealed gain-of-function mutations that target the CD79A and CD79B BCR subunits and the Toll-like receptor signalling adaptor MYD88 [28, 30], with MYD88(L265P) being the most prevalent isoform.

The most striking response rate (80%) was observed in tumours with both CD79B and MYD88(L265P) mutations, but how these mutations cooperate to promote dependence on BCR signalling remains unclear.

We discovered a new mode of oncogenic BCR signalling in ibrutinib-responsive cell lines and biopsies, coordinated by a multiprotein supercomplex formed by MYD88, TLR9 and the BCR (hereafter termed the My-T-BCR supercomplex).

The My-T-BCR supercomplex co-localizes with mTOR on endolysosomes, where it drives pro-survival NF-κB and mTOR signalling.

Inhibitors of BCR and mTOR signalling cooperatively decreased the formation and function of the My-T-BCR supercomplex, providing mechanistic insight into their synergistic toxicity for My-T-BCR⁺ DLBCL cells.

Main

The survival of BCR-dependent ABC lines relied on Toll-like receptor (TLR9), which coordinates MYD88 signalling in innate immune cells, and on two chaperones that regulate the subcellular localization of TLR9, CNPY3 and UNC93B1.

To identify additional components of the My-T-BCR supercomplex, we expressed a MYD88(L265P)–BioID2 protein in three ABC lines and performed mass spectrometry analysis of MYD88-proximal biotinylated proteins.

The My-T-BCR supercomplex coordinates pro-survival signalling in ABC DLBCL, we hypothesized that inhibition of BTK activity by ibrutinib might disrupt this signalling complex.

To globally assess the effect of ibrutinib on the My-T-BCR supercomplex, we treated two ABC lines bearing MYD88(L265P)–BioID2 with ibrutinib and analysed the biotinylated proteins by mass spectrometry.

We provide genetic, proteomic, cell biological and functional evidence for a pro-survival signalling hub—termed the My-T-BCR supercomplex—that coordinates NF-κB activation in DLBCL and identifies tumours that respond to therapeutic inhibition of NF-κB by ibrutinib.

Methods

When surface proteins were targeted, knockout was validated by flow cytometry by spinning cells down, washing in FACS buffer (PBS plus 2% (v/v) FBS, 1 mM

EDTA), and stained at 4 °C for 30 min in FACS buffer with fluorescently labelled antibodies: mouse anti-human CD19-APC (Biolegend SJ25C1, 1:500), mouse anti-human CD81-PE (Biolegend 5A6, 1:500), mouse anti-human IgM-APC (MHM-88), 1:400); or from Southern Biotech: goat anti-human IgG-PE (1:200).

DLBCL cell lines were left untreated, treated with 10 nM ibrutinib, 200 nM AZD2014 or equivalent volumes of DMSO, or transduced with control shRNA (SC4) or shRNAs targeting CD79A, TLR9, MYD88, CARD11, BCL10 or MALT1, followed by puromycin (Invitrogen) selection as previously described [30].

Acknowledgements

A machine generated summary based on the work of Phelan, James D.; Young, Ryan M.; Webster, Daniel E.; Roulland, Sandrine; Wright, George W.; Kasbekar, Monica; Shaffer, Arthur L.; Ceribelli, Michele; Wang, James Q.; Schmitz, Roland; Nakagawa, Masao; Bachy, Emmanuel; Huang, Da Wei; Ji, Yanlong; Chen, Lu; Yang, Yandan; Zhao, Hong; Yu, Xin; Xu, Weihong; Palisoc, Maryknoll M.; Valadez, Racquel R.; Davies-Hill, Theresa; Wilson, Wyndham H.; Chan, Wing C.; Jaffe, Elaine S.; Gascoyne, Randy D.; Campo, Elias; Rosenwald, Andreas; Ott, German; Delabie, Jan; Rimsza, Lisa M.; Rodriguez, Fausto J.; Estephan, Fayez; Holdhoff, Matthias; Kruhlak, Michael J.; Hewitt, Stephen M.; Thomas, Craig J.; Pittaluga, Stefania; Oellerich, Thomas; Staudt, Louis M. 2018 in Nature.

The shieldin complex mediates 53BP1-dependent DNA repair

https://doi.org/10.1038/s41586-018-0340-7

Abstract-Summary

53BP1 is a chromatin-binding protein that regulates the repair of DNA double-strand breaks by suppressing the nucleolytic resection of DNA termini [31, 32].

How 53BP1-pathway proteins shield DNA ends is currently unknown, but there are two models that provide the best potential explanation of their action.

In one model the 53BP1 complex strengthens the nucleosomal barrier to end-resection nucleases [33, 34], and in the other 53BP1 recruits effector proteins with end-protection activity.

Shieldin localizes to double-strand-break sites in a 53BP1- and RIF1-dependent manner, and its SHLD2 subunit binds to single-stranded DNA via OB-fold domains that are analogous to those of RPA1 and POT1.

We show that binding of single-stranded DNA by SHLD2 is critical for shieldin function, consistent with a model in which shieldin protects DNA ends to mediate 53BP1-dependent DNA repair.

Main

To discover proteins acting in the 53BP1 pathway, we searched for genes whose mutation restores homologous recombination in BRCA1-deficient cells and leads to resistance to poly(ADP-ribose) polymerase (PARP) inhibition, which is a hallmark of 53BP1 deficiency [35–37].

For reasons that will become apparent, we named this complex shieldin and renamed C20orf196, FAM35A and CTC-534A2.2 as SHLD1, SHLD2 and SHLD3, respectively.

We conclude that shieldin loss causes resistance to PARPi in both human and mouse BRCA1-deficient tumour cells by reactivating homologous recombination.

These results suggest that SHLD2 mediates 53BP1-dependent DNA repair.

Because the m1 mutation produces a protein that is defective both in ssDNA-binding and suppression of homologous recombination, but which is proficient in both complex assembly and DSB recruitment, we conclude that ssDNA binding by shieldin is critical for 53BP1-dependent DSB repair.

Methods

Twenty-four hours after transfection, cells were selected for 24–48 h with 15 μg/ml puromycin, followed by single clone isolation.

Efficiency of indel formation was analysed by performing PCR amplification of the region surrounding the sgRNA sequence and TIDE analysis on DNA isolated from GFP-expressing cells 9 d post-transduction.

For V5–CTC534A2.2 immunoprecipitations, lysates from one confluent 10-cm dish of 293 T cells transfected with 10 μg pcDNA5.1-FRT/TO-V5-CTC534A2.2 vector was incubated with 10 μg/ml anti-V5 antibody (Invitrogen) for 2 h at 4 °C.

Forty-eight hours after transfection, or 24 h after 0.5 μg/ml doxycyclin induction, cells were treated with 5 Gy X-ray irradiation or micro-irradiated, pre-extracted 10 min on ice with NuEx buffer (20 mM HEPES, pH 7.4, 20 mM NaCl, 5 mM MgCl2, 0.5% NP-40, 1 mM DTT and protease inhibitors) followed by 10 min 2% PFA fixation 1 h post- ionizing radiation or micro-irradiation.

Acknowledgements

A machine generated summary based on the work of Noordermeer, Sylvie M.; Adam, Salomé; Setiaputra, Dheva; Barazas, Marco; Pettitt, Stephen J.; Ling, Alexandra K.; Olivieri, Michele; Álvarez-Quilón, Alejandro; Moatti, Nathalie; Zimmermann, Michal; Annunziato, Stefano; Krastev, Dragomir B.; Song, Feifei; Brandsma, Inger; Frankum, Jessica; Brough, Rachel; Sherker, Alana; Landry, Sébastien; Szilard, Rachel K.; Munro, Meagan M.; McEwan, Andrea; Goullet de Rugy, Théo; Lin, Zhen-Yuan; Hart, Traver; Moffat, Jason; Gingras, Anne-Claude; Martin, Alberto; van Attikum, Haico; Jonkers, Jos; Lord, Christopher J.; Rottenberg, Sven; Durocher, Daniel 2018 in Nature.

High prevalence of Streptococcus pyogenes Cas9-reactive T cells within the adult human population

https://doi.org/10.1038/s41591-018-0204-6

Abstract-Summary

S. pyogenes is a common cause for infectious diseases in humans, but it remains unclear whether it induces a T cell memory against the Cas9 nuclease [38, 39].

We show the presence of a preexisting ubiquitous effector T cell response directed toward the most widely used Cas9 homolog from S. pyogenes (SpCas9) within healthy humans.

We characterize SpCas9-reactive T cells within the CD4/CD8 compartments for multi-effector potency, cytotoxicity, and lineage determination.

In-depth analysis of SpCas9-reactive T cells reveals a high frequency of SpCas9-reactive regulatory T cells that can mitigate SpCas9-reactive effector T cell proliferation and function in vitro.

Main

Intracellular protein degradation processes lead to peptide presentation of Cas9 fragments on the cellular surface of gene-edited cells that may be recognized by SpCas9-reactive T cells.

Our findings indicate that SpCas9-reactive T_{reg} cells are an inherent part of the physiological SpCas9-induced T cell response in the peripheral blood of adult humans.

This could indicate that donors with fewer SpCas9-reactive T_{reg} cells have relatively higher CD8^{+} T_{eff} responses.

CRISPR–Cas9 gene-edited cells will be targeted by the preprimed T_{eff} response directed toward SpCas9.

Our in vitro data indicate that endogenous SpCas9-reactive T_{reg} cells have the potential to mitigate the activation, expansion, and function of SpCas9-reactive T_{eff}, but future studies need to elucidate whether they harbor SpCas9-specific TCRs.

Adoptive transfer of ex vivo enriched SpCas9-reactive T_{reg} cells or T_{reg} cells genetically modified to express SpCas9-specific TCRs may be a therapeutic option.

What might be the physiological significance of SpCas9-reactive T_{reg} cells?

Methods

We incubated them in the presence of 5 µg ml^{-1} SpCas9 whole protein and 1 µg ml^{-1} CD40-specific antibody (HB 14; Miltenyi Biotech) at cell concentrations of 1×10^{7} PBMCs per 1 ml very-low-endotoxin-RPMI 1640 medium with stable glutamine supplemented with 100 U ml^{-1} penicillin, 0.1 mg ml^{-1} streptomycin and 5% heat-inactivated human AB serum (PAA Laboratories) in polystyrene flat-bottom 24-well plates (Falcon; Corning) at 37 °C and 5% CO_2 in humidified incubators for 16 h. After stimulation, cells were washed with PBS (0.5% BSA) and stained with fluorochrome-conjugated antibodies specific for: CD3 (BV650, clone OKT3); CD4 (PerCP-Cy5.5, clone SK3); CD25 (APC, clone 2A3); CD127 (APC Alexa Fluor 700, clone R34.34; Beckman Coulter); CD137 (PE/Cy7, clone 4B4-4); and CD154 (BV711, clone 24–31) for 10 min.

Acknowledgements

A machine generated summary based on the work of Wagner, Dimitrios L.; Amini, Leila; Wendering, Desiree J.; Burkhardt, Lisa-Marie; Akyüz, Levent; Reinke, Petra; Volk, Hans-Dieter; Schmueck-Henneresse, Michael 2018 in Nature Medicine.

CRISPR/Cas9-mediated PD-1 disruption enhances human mesothelin-targeted CAR T cell effector functions.

https://doi.org/10.1007/s00262-018-2281-2

Abstract-Summary

The interaction between programmed cell death protein 1 (PD-1) on activated T cells and its ligands on a target tumour may limit the capacity of chimeric antigen receptor (CAR) T cells to eradicate solid tumours.

PD-1 blockade could potentially enhance CAR T cell function.

To overcome the suppressive effect of PD-1 on CAR T cells, we utilized CRISPR/Cas9 ribonucleoprotein-mediated editing to disrupt the programmed cell death-1 (PD-1) gene locus in human primary T cells, resulting in a significantly reduced PD-1^{hi} population.

CAR T cells with PD-1 disruption show enhanced tumour control and relapse prevention in vivo when compared with CAR T cells with or without αPD-1 antibody blockade.

Extended:

CAR T cells with a 4-1BB signalling domain showed the greatest persistence in a tumour model [15] and advanced toward towards a central memory phenotype [40].

CAR T cells have been reported as hypofunctional in the tumour microenvironment [41] partially due to the suppressive effect of their own expression of PD-1.

We show that CAR T cells that specifically recognize Meso can target Meso-expressing tumours (TNBC) and that this effect is further enhanced by combination with CRISPR/Cas9-mediated PD-1 genome modification.

Introduction

The impressive clinical outcomes associated with the adoptive transfer of CAR-expressing T (CAR T) cells in B-lineage malignancy therapy has prompted its application in the treatment of solid tumours.

Tchou and others have previously shown that Meso CAR T cells display cytotoxic activity towards TNBC cell lines highly expressing Meso [42], but their effector function requires further characterization.

The combination of PD-1 disruption through CRISPR/Cas9 and CD19-CAR T cell therapy against B cell lymphoma and other models has been studied [43–45], but whether PD-1 disruption would improve CAR T cell therapy in a breast cancer model has not been fully reported in the literature.

We show that CAR T cells that specifically recognize Meso can target Meso-expressing tumours (TNBC) and that this effect is further enhanced by combination with CRISPR/Cas9-mediated PD-1 genome modification.

Materials and methods

The following antibodies were used for cell surface and intracellular staining: APC-PD1 (eBioscience, clone J105), FITC-Meso (R&D, clone 420411), FITC-CD4 (BD Biosciences, clone RPA-T4), PE-CD25 (BD Biosciences, clone 4E3), PerCP-CD8 (BD Biosciences, clone RPA-T8), PE-IL-2 (BD Biosciences, clone MQ1-17H12), APC-IFNγ (BD Biosciences, clone B27), CD3 (BD Biosciences, clone UCH71), and PE-CD107a (BD Biosciences, clone H4A3) antibodies.

At day 5 after activation, PD-1 sgRNA was electroporated into T cells with a Neon transfection kit and device (Thermo Fisher Scientific).

At day 14, T cells were stimulated by Dynabeads® Human T-Activator CD3/CD28 (Thermo Fisher Scientific) and cultured in a 37 °C incubator for 48 h. Then, PD-1 and CD25 staining was performed.

Peripheral blood was drawn from the tail vein and stained with anti-human CD45 antibody (clone H130, Biolegend) at the indicated experimental time points and mixed with CountBright™ Absolute Counting Beads (Thermo Fisher Scientific) as an event collecting control.

Results

Even with the 4-1BB signal, which supports a moderate rise in the CD8 T cell fraction, the Meso CAR T cell groups (with or without PD-1 sgRNA-Cas9 RNP) showed lower CD8 counts than the control group, which suggests that the electroporation procedure might affect the growth of CD8 T cells more than that of CD4 T cells.

Expression of PD-L1 in tumour cells could be further stimulated (without significant differences) when the tumour cells were co-cultured with Meso CAR/PD-1 sgRNA-Cas9 RNP T cells.

We next examined the cytokine production and cytotoxic capacities of Meso CAR/PD-1 sgRNA-Cas9 RNP T cells.

PD-1 disruption by sgRNA-Cas9 RNP exhibited significantly higher potency in enhancing cytotoxicity and the production of cytokines, IL-2 and IFNγ by Meso CAR T cells when compared that by antibody blockade.

We observed significantly greater hCD3[+] T cell infiltration in the tumour of Meso CAR/PD-1 sgRNA-Cas9 RNP T cell group than in that of the Meso CAR T cell group.

Discussion

Reports have shown that CRISPR/Cas9-mediated disruption of PD-1 could boost the function of CD19-CAR T cells [44, 46] and that Epstein–Barr virus LMP2 peptide increased the cytotoxic function of T lymphocytes to different degrees [45].

Upon PD-1 disruption, Meso CAR T cells showed increased cytotoxicity and cytokine production but not increased proliferation in vitro.

Our data suggested that PD-1 disruption through CRISPR/Cas9 could not only enhance tumour control by Meso CAR T cells but also delay tumour relapse to some degree after CAR T cell treatment.

We showed that PD-1 disruption mediated by CRISPR/Cas9 improved the effect of the CAR, which contains 4-1BB and CD3 (BBz) as signalling domains, on T cell function.

Although Meso 28z and BBz-CAR T cells exhibited equivalent secretion of effector cytokines and proliferation in vitro upon initial antigen stimulation, Meso 28z-CAR T cells expressed higher levels of exhaustion markers, such as PD-1, TIM-3, and LAG-3, upon further stimulation than Meso BBz-CAR T cells [47].

Acknowledgements

A machine generated summary based on the work of Hu, Wanghong; Zi, Zhenguo; Jin, Yanling; Li, Gaoxin; Shao, Kang; Cai, Qiliang; Ma, Xiaojing; Wei, Fang 2018 in Cancer Immunology, Immunotherapy.

Identification of preexisting adaptive immunity to Cas9 proteins in humans.

https://doi.org/10.1038/s41591-018-0326-x

Abstract-Summary

These two bacterial species infect the human population at high frequencies [48, 49], we hypothesized that humans may harbor preexisting adaptive immune responses to the Cas9 orthologs derived from these bacterial species, SaCas9 (S. aureus) and SpCas9 (S. pyogenes).

We also found anti-SaCas9 T cells in 78% and anti-SpCas9 T cells in 67% of donors, which demonstrates a high prevalence of antigen-specific T cells against both orthologs.

We confirmed that these T cells were Cas9-specific by demonstrating a Cas9-specific cytokine response following isolation, expansion, and antigen restimulation.

These data demonstrate that there are preexisting humoral and cell-mediated adaptive immune responses to Cas9 in humans, a finding that should be taken into account as the CRISPR–Cas9 system moves toward clinical trials.

Main

The abundance of S. aureus and S. pyogenes within the human population, as well as widespread adaptive humoral and cell-mediated immune responses to both species, raises the possibility that humans may also have preexisting adaptive immunity to the Cas9 orthologs derived from these bacteria.

To test whether a preexisting immune response to Cas9 also existed, we used three different highly sensitive methods to detect antigen-reactive T cells: interferon-γ (IFN-γ) enzyme-linked immunospot (ELISpot); intracellular cytokine staining (ICS); and detection of activated T cells by the expression of CD137 or CD154 on the cell surface [50–53].

We detected IgG antibodies against both SaCas9 and SpCas9 in the study population, and could also detect antigen-reactive T cells directed against both Cas9 orthologs.

Preexisting adaptive immune responses to Cas9 may not prove to be a major barrier to the implementation of ex vivo therapies, which involve the use of Cas9 to edit cells outside of direct contact with the human immune system.

Methods

After 2 h of incubation with antigen, brefeldin A (Abcam) was added at a concentration of 10 ug ml^{-1} and cells were incubated for another 4 h. Cells were washed two times in staining buffer (1% human AB serum and 0.02% sodium azide in PBS, pH7) and surface-stained for CD3 (PerCP/Cy5.5 anti-human CD3 Antibody, clone HIT3a; BioLegend), CD4 (Brilliant Violet 605 anti-human CD4 antibody, clone OKT4; BioLegend), and CD8a (Brilliant Violet 421 anti-human CD8a Antibody, clone RPA-T8; BioLegend).

After an overnight rest, cells were incubated with each antigen at a concentration of 10 ug ml^{-1} for 16 h. Cells were then washed two times in staining buffer and stained with CD3 (PerCP/Cy5.5, clone HIT3a), CD4 (Brilliant Violet 605, clone OKT4), CD8a (Brilliant Violet 421, clone RPA-T8), CD137 (PE/Cy7 anti-human CD137 (4-1BB) Antibody, clone 4B4-1; BioLegend), and CD154 (PE anti-human CD154 Antibody, clone 24–31; BioLegend).

Acknowledgements

A machine generated summary based on the work of Charlesworth, Carsten T.; Deshpande, Priyanka S.; Dever, Daniel P.; Camarena, Joab; Lemgart, Viktor T.; Cromer, M. Kyle; Vakulskas, Christopher A.; Collingwood, Michael A.; Zhang, Liyang; Bode, Nicole M.; Behlke, Mark A.; Dejene, Beruh; Cieniewicz, Brandon; Romano, Rosa; Lesch, Benjamin J.; Gomez-Ospina, Natalia; Mantri, Sruthi; Pavel-Dinu, Mara; Weinberg, Kenneth I.; Porteus, Matthew H. 2019 in Nature Medicine.

Prioritization of cancer therapeutic targets using CRISPR–Cas9 screens.

https://doi.org/10.1038/s41586-019-1103-9

Abstract-Summary

We integrated cell fitness effects with genomic biomarkers and target tractability for drug development to systematically prioritize new targets in defined tissues and genotypes.

Our analysis provides a resource of cancer dependencies, generates a framework to prioritize cancer drug targets and suggests specific new targets.

The principles described in this study can inform the initial stages of drug development by contributing to a new, diverse and more effective portfolio of cancer drug targets.

Main

The molecular features of a patient's tumour influence clinical responses and can be used to guide therapy, leading to more effective treatments and reduced toxicity [54].

Unbiased strategies that effectively identify and prioritize targets in tumours could expand the range of targets, improve success rates and accelerate the development of new cancer therapies.

Defining core and context-specific fitness genes

Fitness genes that are common to the majority of tested cell lines or common within a cancer type (referred to as pan-cancer or cancer-type-specific core fitness genes, respectively) may be involved in essential processes in cells and have greater toxicity.

To previously identified reference core fitness gene sets [55, 56], our pan-cancer core fitness gene set showed greater recall of genes involved in essential processes (median = 67%, versus 28% and 51% in the previously published gene sets of refs.

A quantitative framework for target prioritization

For each gene, 70% of the priority score was derived from CRISPR–Cas9 experimental evidence and averaged across dependent cell lines on the basis of the fitness effect size, the significance of fitness deficiency, target gene expression, target mutational status and evidence for other fitness genes in the same pathway.

The remaining 30% of the priority score was based on evidence of a genetic biomarker that was associated with a target dependency and the frequency at which the target was somatically altered in tumours in patients [57].

Tractability assessment of priority targets

Targets in group 2 are most likely to be novel and tractable through conventional modalities and, therefore, represent good candidates for drug development.

Our framework informed a data-driven list of prioritized therapeutic targets that would be strong candidates for the development of cancer drugs.

WRN is a target in cancers with MSI

WRN is one of five RecQ family DNA helicases, of which it is the only one that has both a helicase and an exonuclease domain, and has diverse roles in DNA repair, replication, transcription and telomere maintenance [58].

Other tested RecQ family members (BLM, RECQL and RECQL5) were not associated as fitness genes in MSI cell lines.

To determine whether the loss-of-fitness effect was selective to WRN and identify a potential strategy for drug targeting, we performed functional rescue experiments using wild-type, or hypomorphic versions of mouse Wrn (resistant to the WRN sgRNAs that we used) with a mutation in the exonuclease (E78A) or helicase (R799C or T1052G) domain to impair protein function [59–61].

Discussion

New approaches are needed to effectively prioritize candidate therapeutic targets for cancer treatments.

We performed CRISPR–Cas9 screens in a diverse collection of cancer cells lines and combined this with genomic and tractability data to systematically nominate new cancer targets in an unbiased way.

Even a modest improvement in drug-development success rates, and an expanded repertoire of targets, through approaches such as ours could provide benefits to patients with cancer.

We identified WRN as a promising new synthetic lethal target in MSI tumours.

Loss of WRN is compatible with human development; however, targeting WRN could result in damage to normal cells.

We developed an unbiased and systematic framework that effectively ranks priority targets, such as WRN.

Methods

Class C marker or weaker: an ANOVA P-value threshold of 10^{-3} and for pan-cancer associations at least one Glass's $\Delta > 1$; for weaker, a simple Student's t-test (for difference assessment of the mean depletion fold change between CDE-positive/CDE-negative cell lines) P-value threshold of 0.05 and for pan-cancer associations, at least one Glass's $\Delta > 1$.

For GPX4 analysis, cell lines were divided into two groups according to their loss-of-fitness response to GPX4 knockout (using BAGEL FDR < 5% as significance threshold for gene depletion) and gene expression fold changes were calculated between the GPX4 non-dependent and dependent cell lines ($\log_2$ values of the mean difference).

At day 0, cells were transduced with viral particles containing sgRNAs targeting essential or non-essential genes, or WRN sgRNA 1 and WRN sgRNA 4 in order to achieve a >90% transduction efficiency.

Acknowledgements

A machine generated summary based on the work of Behan, Fiona M.; Iorio, Francesco; Picco, Gabriele; Gonçalves, Emanuel; Beaver, Charlotte M.; Migliardi, Giorgia; Santos, Rita; Rao, Yanhua; Sassi, Francesco; Pinnelli, Marika; Ansari, Rizwan; Harper, Sarah; Jackson, David Adam; McRae, Rebecca; Pooley, Rachel; Wilkinson, Piers; van der Meer, Dieudonne; Dow, David; Buser-Doepner, Carolyn; Bertotti, Andrea; Trusolino, Livio; Stronach, Euan A.; Saez-Rodriguez, Julio; Yusa, Kosuke; Garnett, Mathew J. 2019 in Nature.

Sequential LASER ART and CRISPR Treatments Eliminate HIV-1 in a Subset of Infected Humanized Mice.

https://doi.org/10.1038/s41467-019-10366-y

Abstract-Summary

Elimination of HIV-1 requires clearance and removal of integrated proviral DNA from infected cells and tissues.

Sequential long-acting slow-effective release antiviral therapy (LASER ART) and CRISPR-Cas9 demonstrate viral clearance in latent infectious reservoirs in HIV-1 infected humanized mice.

HIV-1 subgenomic DNA fragments, spanning the long terminal repeats and the Gag gene, are excised in vivo, resulting in elimination of integrated proviral DNA; virus is not detected in blood, lymphoid tissue, bone marrow and brain by nested and digital-droplet PCR as well as RNAscope tests.

HIV-1 is readily detected following sole LASER ART or CRISPR-Cas9 treatment.

Extended:

In these three independent experiments, one with HIV-1_{NL4-3} and the other two with HIV-1_{ADA} infection of humanized mice, single treatments with LASER ART or AAV$_9$-CRISPR-Cas9 resulted in viral rebound in 100% of treated infected animals.

Introduction

This includes inadequate therapeutic access to viral reservoirs, rapid spread of infection by continuous sources of virus and susceptible cells and a failure to eliminate residual latent integrated proviral DNA.

Multimodal robust pharmaceutic strategies are needed for complete elimination of HIV-1 if no viral resurgence after cessation of ART is to be achieved.

LASER ART alone cannot rid the infected host of latent HIV-1 no matter how successful the drugs may prove to be at restricting viral infection.

The two approaches are combined to examine whether LASER ART and CRISPR-Cas9 treatments could provide combinatorial benefit for viral elimination.

Viral clearance is achieved from HIV-1 infected spleen and lymphoid tissues as well as a broad range of solid organs from documented prior infected humanized mice treated with LASER ART and AAV$_9$-CRISPR-Cas9.

We conclude that viral elimination by a combination of LASER ART and gene editing strategy is possible.

Results

In the group that received LASER ART with subsequent AAV$_9$-CRISPR-Cas9, viral rebound was not observed in two animals.

The efficiency of the proviral DNA excision by CRISPR-Cas9 in the spleens of two infected humanized mice from the CRISPR-Cas9 and LASER ART group (animals where no rebound was observed) was determined by ddPCR.

Splenocytes and bone marrow cells were isolated from HIV-1 infected mice with or without prior LASER ART and/or CRISPR-Cas9 treatments at the study end.

We confirmed the ability of LASER ART and CRISPR-Cas9 to eliminate viral rebound in a new cohort of CD34+ HSC-reconstituted animals infected with HIV-1_{ADA}.

In these three independent experiments, one with HIV-1$_{NL4-3}$ and the other two with HIV-1$_{ADA}$ infection of humanized mice, single treatments with LASER ART or AAV$_9$-CRISPR-Cas9 resulted in viral rebound in 100% of treated infected animals.

Discussion

One may predict that, any or all steps towards HIV elimination must include precise targeted ART delivery, maintenance of vigorous immune control, effective blockade of viral growth and immune-based elimination of pools of infected cells or genome integrated proviral DNA.

The sustained human grafts as confirmed by flow cytometry were viable and functional for more than 6 months, which provided a platform that allowed treatment interventions for prolonged time periods and a clear ability during ART to best establish a continuous latent HIV-1 reservoir in peripheral tissues and the brain and the noted immunological responses to the viral infection [62–67].

The success in these prior studies led to the use of LASER ART in the current report in order to maximize ART ingress to cell and tissue sites of viral replication enabling the drugs to reach these sites at high concentrations for sustained time periods.

Methods

Total cellular DNA obtained from the HIV-1 infected cell line ACH2 served as a positive control and standards, while human genomic DNA obtained from uninfected NSG-hu mice served as a negative control.

Cell-associated HIV-1 RNA and DNA were quantified by real-time qPCR and droplet digital PCR (ddPCR) assays.

Because of extremely low numbers of latently-infected human cells in HIV-infected NSG-hu mice after long-term ART, detection of total HIV-1 DNA, required two rounds of PCR amplification.

PBMCs obtained from leukopaks from HIV-1,2 seronegative donors were stimulated with PHA and IL-2 and co-cultured with human bone marrow or spleen cells recovered from 3 groups of CD34+ HSC- NSG mice that included HIV- 1 infected, infected and LASER ART treated, and LASER ART and AAV$_9$-CRISPR-Cas9-treated mice.

Tissue drug levels, HIV-1 RT activity, HIV-1p24 antigen staining, T-cell populations, viral RNA and DNA, and viral load were analyzed by one-way ANOVA with Bonferroni correction for multiple-comparisons.

Acknowledgements

A machine generated summary based on the work of Dash, Prasanta K.; Kaminski, Rafal; Bella, Ramona; Su, Hang; Mathews, Saumi; Ahooyi, Taha M.; Chen, Chen; Mancuso, Pietro; Sariyer, Rahsan; Ferrante, Pasquale; Donadoni, Martina; Robinson, Jake A.; Sillman, Brady; Lin, Zhiyi; Hilaire, James R.; Banoub, Mary; Elango, Monalisha; Gautam, Nagsen; Mosley, R. Lee; Poluektova, Larisa Y.; McMillan, JoEllyn; Bade, Aditya N.; Gorantla, Santhi; Sariyer, Ilker K.; Burdo,

Tricia H.; Young, Won-Bin; Amini, Shohreh; Gordon, Jennifer; Jacobson, Jeffrey M.; Edagwa, Benson; Khalili, Kamel; Gendelman, Howard E. 2019 in Nature Communications.

The CoQ oxidoreductase FSP1 acts parallel to GPX4 to inhibit ferroptosis.

https://doi.org/10.1038/s41586-019-1705-2

Abstract-Summary

Ferroptosis is a form of regulated cell death that is caused by the iron-dependent peroxidation of lipids [68, 69].

Sensitivity to GPX4 inhibitors varies greatly across cancer cell lines [70], which suggests that additional factors govern resistance to ferroptosis.

Using a synthetic lethal CRISPR–Cas9 screen, we identify ferroptosis suppressor protein 1 (FSP1) (previously known as apoptosis-inducing factor mitochondrial 2 (AIFM2)) as a potent ferroptosis-resistance factor.

We further find that FSP1 expression positively correlates with ferroptosis resistance across hundreds of cancer cell lines, and that FSP1 mediates resistance to ferroptosis in lung cancer cells in culture and in mouse tumour xenografts.

These findings define a ferroptosis suppression pathway and indicate that pharmacological inhibition of FSP1 may provide an effective strategy to sensitize cancer cells to ferroptosis-inducing chemotherapeutic agents.

Main

GPX4 is considered to be the primary enzyme that prevents ferroptosis [69].

The resistance of some cancer cell lines to GPX4 inhibitors [70] led us to search for additional protective pathways.

FSP1 is a potent ferroptosis suppressor

$FSP1^{KO}$ cells displayed increased sensitivity to additional ferroptosis inducers, including the GPX4 inhibitor ML162 and the system x_c^- inhibitor erastin2 (ref.

These findings demonstrate that FSP1 is a strong suppressor of ferroptosis.

Plasma-membrane FSP1 blocks ferroptosis

These results indicate that the myristoylation of FSP1 mediates the recruitment of this protein to lipid droplets and the plasma membrane.

FSP1 reduces CoQ to suppress ferroptosis

FSP1 functions as an NADH-dependent CoQ oxidoreductase in vitro [71].

These data indicate that FSP1 and the CoQ synthesis machinery function in the same pathway to suppress lipid peroxidation and ferroptosis.

FSP1 in cancer ferroptosis resistance

FSP1 is a biomarker of ferroptosis resistance in many types of cancer.

To examine the possibility that the inhibition of FSP1 could be a clinically relevant approach to sensitize tumours to ferroptosis-activating chemotherapies, we used ferroptosis-resistant H460 lung cancer cells in a preclinical tumour xenograft mouse model.

To determine whether the growth of $FSP1^{KO}$ tumours can be inhibited by blocking cystine import, we treated H460 cells with imidazole–ketone–erastin (IKE), a system x_c^- inhibitor that can induce ferroptosis in vivo [72].

Ferroptosis has emerged as a potential cause of cell death in degenerative diseases and as a promising strategy to induce the death of cancer cells that are resistant to other therapies [68, 69, 73].

Our findings indicate that FSP1 expression is important for predicting the efficacy of ferroptosis-inducing drugs in cancers and highlight the potential for FSP1 inhibitors [74] as a strategy to overcome ferroptosis resistance in multiple types of cancer.

Methods

U-2 OS FSP1–HaloTag knock-in lines were generated by cotransfection of U-2 OS T-Rex Flp-In cells with the donor plasmid pUC57 (described in 'Plasmids') and PX459 encoding FSP1 sgRNA guide 3 at a 2:1 w/w ratio in medium containing 1 µM SCR7 non-homologous end joining inhibitor (Xcess Biosciences) for 48 h, followed by selection in medium containing 1 µg/ml puromycin.

Ten 15-cm plates of U-2 OS cells expressing inducible FSP1–GFP were induced with 10 ng/ml doxycycline for 48 h. Cells were collected by scraping into PBS and centrifuged for 10 min at 500 g. Cell pellets were resuspended in cold hypotonic lysis medium (HLM, 20 mM Tris–HCl pH 7.4 and 1 mM EDTA) supplemented with 1 × cOmplete, Mini, EDTA-free Protease Inhibitor Cocktail (Sigma-Aldrich), incubated on ice for 10 min, dounced using 80 × strokes and centrifuged at 1000 g for 10 min.

Acknowledgements

A machine generated summary based on the work of Bersuker, Kirill; Hendricks, Joseph M.; Li, Zhipeng; Magtanong, Leslie; Ford, Breanna; Tang, Peter H.; Roberts, Melissa A.; Tong, Bingqi; Maimone, Thomas J.; Zoncu, Roberto; Bassik, Michael C.; Nomura, Daniel K.; Dixon, Scott J.; Olzmann, James A. 2019 in Nature.

Somatic inflammatory gene mutations in human ulcerative colitis epithelium.

https://doi.org/10.1038/s41586-019-1844-5

Abstract-Summary

With ageing, normal human tissues experience an expansion of somatic clones that carry cancer mutations [75–81].

Using whole-exome sequencing data from 76 clonal human colon organoids, we identify a unique pattern of somatic mutagenesis in the inflamed epithelium of patients with ulcerative colitis.

The affected epithelium accumulates somatic mutations in multiple genes that are related to IL-17 signalling—including NFKBIZ, ZC3H12A and PIGR, which are genes that are rarely affected in colon cancer.

Unbiased CRISPR-based knockout screening in colon organoids reveals that the mutations confer resistance to the pro-apoptotic response that is induced by IL-17A. Some of these genetic mutations are known to exacerbate experimental colitis in mice [82–85], and somatic mutagenesis in human colon epithelium may be causally linked to the inflammatory process.

Extended:

Our findings provide a blueprint for how clones that are adapted to an inflammatory microenvironment emerge and expand in a normal tissue, and will prompt future research on the role of such clones in the pathology of human diseases.

Main

During repetitive cycles of self-renewal, somatic cells inevitably acquire genetic mutations [80].

Sequencing efforts have identified such somatic mutations in normal tissues at a clonal or near-clonal level [75–81].

Genetic perturbation studies also support the concept that the acquisition of tissue-specific somatic mutations, or 'somatic evolution', confers cell-autonomous self-renewal capability and contributes to tumorigenesis.

Given the importance of adaptation that is driven by the environment in Darwinian evolution, somatic evolution may also occur in a non-cell-autonomous manner as a consequence of continuous pressure from the hostile tissue environment.

To resolve this issue, we here obtain a panoramic view of the genetic mutations in the UC epithelium using an organoid-based analysis of clonal somatic mutations.

Somatic mutations in the UC epithelium

To determine the mutagenic effect of chronic inflammation on normal colorectal epithelium, we established non-neoplastic colorectal organoids from patients with UC, including patients with colitis-associated neoplasia (CAN), and from healthy control individuals.

To capture inflammation-related mutations at the clonal level, we subsequently expanded clonal organoids from single cells from patients with UC and healthy control individuals, and subjected the organoids to whole-exome sequencing.

Because the mutational burden in tissues affected by UC was expected to reflect cumulative inflammation rather than disease activity at the point of sampling, we divided UC^{inf} organoids on the basis of CAN concomitance to designate organoids with a history of long-term active inflammation.

We next determined whether UC inflammation is related to specific mutational signatures [86].

Despite being derived from a non-neoplastic tissue, UC08 organoids contained an $MLH1^{T117M}$ mutation and showed mutational signatures that are associated with a deficiency in mismatch repair.

Genomic landscape of the UC epithelium

To investigate whether functional genetic mutations were enriched in UC^{inf} organoids, we focused on truncating mutations, including nonsense, frame-shift and splice site mutations.

These results indicate that the UC epithelium is prone to genomic instability, and that this eventually contributes to the formation of a unique genomic landscape.

Recurrent IL-17 pathway mutations in UC

Previous studies have revealed that PIGR is upregulated by IL-17-related activation of NF-κB [87, 88].

These results suggest that IL-17 signal activation upregulates the expression of IκBζ and PIGR in the UC epithelium.

The mutational spectrum in UC epithelium

In three UC^{inf} organoids, we identified five IL17RA mutations that were not detected by whole-exome sequencing.

Multiple disparate truncating mutations often co-existed in a tiny biopsy sample, with PIGR, ARID1A and NFKBIZ mutations being prevalent.

On the basis of the high frequency of mutations in PIGR, we next examined the expression of PIGR in the colonic tissues adjacent to the sampled sites.

NFKBIZ, ZC3H12A and PIGR could be upregulated by cytokines other than IL-17A, we comprehensively examined mutations in cytokine and pathogen-associated molecular pattern (PAMP) receptor genes by targeted sequencing.

Of the genes with truncating mutations, OSMR, IL15RA and TLR3 were found to be expressed at substantial levels in the colonic epithelium [89].

IL-17A cytotoxicity in colonic epithelium

PIGR sgRNAs were not enriched, which suggests that growth selection by PIGR mutations was not mediated directly by IL-17A cytotoxicity but requires an IgA-related immune response.

These results together suggest that IL-17A is detrimental for the epithelial lining, and that clones with IL-17A pathway mutations are able to avert IL-17A-mediated cytotoxicity and thereby selectively expand in tissue that is affected by UC.

Discussion

IL-17A elicited a pro-apoptotic response in normal colonic organoids, and loss-of-function mutations in positive regulators of IL-17 signalling—namely NFKBIZ, IL17RA and TRAF3IP2—were enriched in the UC epithelium.

Although IL-17A induced the expression of PIGR in organoids, high levels of PIGR were present in the colonic epithelium independently of IL-17A stimulation, suggesting that there is room for debate on the convergence of IL-17 signalling mutations.

With the reduction of the IL-17A-induced pro-apoptotic response that was observed here in human organoids with UC-related mutations, loss-of-function mutations of Il17ra [82], Pigr [83] or Traf3ip2 [84] (and gain-of-function of Zc3h12a [85]) exacerbate experimental colitis in genetically engineered mice.

Some of the somatically mutated genes in the UC epithelium are also associated with susceptibility to UC—including NFKBIZ (rs616597), TRAF3IP2 (rs3851228), PIGR (rs3024505) and NOS2 (rs2945412) [90]—which suggests that altered function of these genes may also contribute to the onset of UC.

Methods

Crypt isolation and organoid culture from samples of human large intestine were carried out as previously described [91, 92] following different procedures depending on sample types.

For analysis of numbers of mutations, raw variants were filtered by removing SNVs that were detected in the paired blood samples with VAF > 0.03.

UC06_IS were also excluded owing to the use of different versions of exon-capture kits between organoids and reference blood samples.

Three independent normal colonic organoid lines that were established previously as counterparts of colon cancer organoids [92] were used for gene expression microarray analysis.

Gene set enrichment analysis was performed using this IL-17A target gene set and public datasets with data on gene expression in UC-affected mucosa (GSE38713 [93] and GSE59071 [94]).

Raw variants were filtered by removing variants registered in HGVD and common variants detected in the pooled targeted sequencing data of four blood samples in this study.

Acknowledgements

A machine generated summary based on the work of Nanki, Kosaku; Fujii, Masayuki; Shimokawa, Mariko; Matano, Mami; Nishikori, Shingo; Date, Shoichi; Takano, Ai; Toshimitsu, Kohta; Ohta, Yuki; Takahashi, Sirirat; Sugimoto, Shinya; Ishimaru, Kazuhiro; Kawasaki, Kenta; Nagai, Yoko; Ishii, Ryota; Yoshida, Kosuke; Sasaki, Nobuo; Hibi, Toshifumi; Ishihara, Soichiro; Kanai, Takanori; Sato, Toshiro 2019 in Nature.

Genome-wide CRISPR screen identifies host dependency factors for influenza A virus infection.

https://doi.org/10.1038/s41467-019-13965-x

Abstract-Summary

Host dependency factors that are required for influenza A virus infection may serve as therapeutic targets as the virus is less likely to bypass them under drug-mediated selection pressure.

We perform a genome-wide CRISPR/Cas9 screen and devise a new approach, meta-analysis by information content (MAIC) to systematically combine our results with prior evidence for influenza host factors.

We validate the host factors, WDR7, CCDC115 and TMEM199, demonstrating that these genes are essential for viral entry and regulation of V-type ATPase assembly.

Extended:

Our MAIC algorithm consolidates data from all previous genetic screens and proteomics studies and generates an annotated list of IAV HDFs which can serve as useful resource for future studies.

Introduction

Identification of host dependency factors (HDFs) that are necessary for IAV replication thus provides an attractive strategy for discovering new therapeutic targets, since the evolution of resistance to host-targeted therapeutics is expected to be slower [95–97].

A recently published genome-wide CRISPR/Cas9 screen based on cell survival after IAV infection uncovered a number of new HDFs involved in early IAV infection, but shared few hits with previous RNAi screens.

To more comprehensively identify IAV-host interactions, we perform pooled genome-wide CRISPR/Cas9 screens and use IAV hemagglutinin (HA) protein expression on the cell surface as a phenotypic readout.

We identify an extensive list of IAV HDFs, including new and previously known factors, involved in various stages of the IAV life cycle.

MAIC performs better than other algorithms for both synthetic data and in an experimental test, and provides a comprehensive ranked list of host genes necessary for IAV infection.

Results

To test if the four genes are required for efficient virus production, we infected WDR7, CCDC115, TMEM199, and CMTR1 polyclonal KO cells with H1N1 PR8 virus and H3N2 Udorn virus at MOI 0.1 and monitored virus production at 24, 48, and 72 h post-infection by plaque assay.

All four polyclonal KO cell lines had comparable GFP expression to wild type cells when infected with an MLV-GFP retrovirus pseudotyped with amphotropic MLV-envelope protein, suggesting that WDR7, CCDC115 and TMEM199 are specifically required for IAV entry in a HA/NA dependent manner.

To test if these genes are required for IAV entry by regulating endo-lysosomal acidification, we stained WDR7, CCDC115, TMEM199, and CMTR1 polyclonal KO cells with lysotracker red, fluorescent-labeled anti-Rab7 and anti-LAMP1 antibodies.

Discussion

Three of our top ranked hits from the CRISPR screens, WDR7, CCDC115 and TMEM199, have been reported as putative V-type ATPase-associated co-factors [98–101], but their functions in mammalian cells and especially in the context of viral infections are poorly understood.

The unexpected observation that WDR7, CCDC115, and TMEM199 polyclonal KO cells underwent expansion and over-acidification of the endo-lysosomal compartments led us to hypothesize that long-term inhibition of V-type ATPases may cause a compensatory increase in lysosomal function, a phenomenon that is observed in cells that were subjected to starvation or prolonged treatment with lysosomoptropic compounds [102, 103].

While it has been established that an acidic endosomal environment is required for IAV entry [104, 105], we showed that depletion of WDR7, CCDC115, and TMEM199 increases endo-lysosomal acidification, yet reduces viral infection.

Unlike a previous study which showed that siRNA knockdown of CMTR1 causes up-regulation of IFN-β in the absence of additional stimulation [106], we found differential expression of type I IFN and IFN-stimulated genes (ISGs) only when the cells were infected with IAV.

Methods

On day 9 post-transduction, 200–400 million puromycin resistant A459-Cas9 cells were infected with Influenza A PR8 virus at MOI5 for 16 h. They were then washed and stained with florescent anti-Influenza A HA (AB1074) antibody.

GFP+ cells were infected with influenza A virus for 16 h at MOI5 and stained for surface HA using anti-influenza A HA antibody (AB1074).

A549 or NHLFs cells were infected with Influenza A PR8 or Udorn virus at MOI 0.1 in serum-free DMEM supplemented with 1% BSA and 1 μg/μl TPCK trypsin.

A549 cells were transduced with pLentiCRISPR-V2 expressing sgRNA against genes of interest and selected with 1 μg/μl Puromycin for 2 days.

The cells were infected with Influenza A PR8 virus the following day for 4 h at 37°.

Infected cells were then fixed with 4% paraformaldehyde and stained with FITC anti-influenza A NP antibody overnight at 4°.

Acknowledgements

A machine generated summary based on the work of Li, Bo; Clohisey, Sara M.; Chia, Bing Shao; Wang, Bo; Cui, Ang; Eisenhaure, Thomas; Schweitzer, Lawrence D.; Hoover, Paul; Parkinson, Nicholas J.; Nachshon, Aharon; Smith, Nikki; Regan, Tim; Farr, David; Gutmann, Michael U.; Bukhari, Syed Irfan; Law, Andrew; Sangesland, Maya; Gat-Viks, Irit; Digard, Paul; Vasudevan, Shobha; Lingwood, Daniel;

Dockrell, David H.; Doench, John G.; Baillie, J. Kenneth; Hacohen, Nir 2020 in Nature Communications.

Genome-wide CRISPR–Cas9 screening reveals ubiquitous T cell cancer targeting via the monomorphic MHC class I-related protein MR1.

https://doi.org/10.1038/s41590-019-0578-8

Abstract-Summary

Human leukocyte antigen (HLA)-independent, T cell–mediated targeting of cancer cells would allow immune destruction of malignancies in all individuals.

We use genome-wide CRISPR–Cas9 screening to establish that a T cell receptor (TCR) recognized and killed most human cancer types via the monomorphic MHC class I-related protein, MR1, while remaining inert to noncancerous cells.

Unlike mucosal-associated invariant T cells, recognition of target cells by the TCR was independent of bacterial loading.

Concentration-dependent addition of vitamin B-related metabolite ligands of MR1 reduced TCR recognition of cancer cells, suggesting that recognition occurred via sensing of the cancer metabolome.

Main

Established unconventional T cell ligands include lipid antigens presented by the conserved CD1 family of molecules, as recognized by natural killer T (NKT) cells and germline-encoded mycolyl-lipid reactive T (GEM) cells.

In combination, these data suggest that MR1 presents a wide range of metabolic intermediates at the cell surface, in much the same way that MHC molecules present arrays of peptides, and the CD1 family of molecules present various lipid antigens.

These MR1-restricted T cells are not classical MAIT cells, in that they do not appear to possess a TRAV1-2 TCR, nor do they react to bacterial antigens bound to MR1.

We report a non-MAIT TCR that recognizes a nonbacterial antigen restricted by MR1, which results in lysis of cancer cells.

Results

Combined with the fact that β2M and MR1 heterodimerize to form a monomorphic stable antigen-presenting molecule that is known to activate MAITs and other MR1-restricted T cells, these data strongly suggested that the MC.7.G5 T cell recognized cancer targets via the MR1 molecule.

The requirement for ligand-binding K43 suggested that MC.7.G5 might recognize an MR1-bound ligand that was specifically expressed or upregulated in malignant cells.

MC.7.G5 exhibited cancer specificity, unlike the majority of MR1T cells [107], which require overexpression of MR1 for optimal target-cell recognition and also are activated in response to MR1 expression by healthy monocyte-derived dendritic cells.

These results indicate that MC.7.G5 does not exclusively recognize MR1 per se, nor recognize MR1 by known mechanisms, but rather it recognizes MR1 with bound cargo that is specific to, or associated with, cancer cells.

Discussion

The level of MR1 surface expression required for cancer cell recognition by MC.7.G5 was often below the threshold required for staining with antibody, suggesting that the MC.7.G5 TCR might be capable of responding to a low copy number of the MR1 ligand, which is akin to T cells that recognize pMHC and MAIT TCR recognition of MR1 (ref [108]).

Knowledge of known MR1-restricted ligands suggests that the MC.7.G5 TCR ligand may be a cancer-specific or -associated metabolite.

This suggests that the MR1-associated ligand targeted by the MC.7.G5 TCR is part of a pathway essential for the basic survival of cancer cells, and therefore not amenable to the gene disruption required for CRISPR–Cas9 screening.

Discovery of MR1-restricted ligands recognized by MC.7.G5-like T cells may further open up opportunities for therapeutic vaccination for many cancers in all individuals.

Methods

Healthy T and B cells were purified from PBMCs using CD3 (negative purification) or CD19 magnetic beads (Miltenyi Biotec), then activated with 1 µg ml^{-1} phyto-hemagglutinin or 1 µM of TLR9 ligand ODN 2006 (Miltenyi Biotec), respectively, for 24 h. Mouse anti-human CD69 (clone FN50, BioLegend) was used to confirm activation.

Cells were collected from culture then incubated with 100–200 µM of tBHP or H_2O_2 for 1 h in R10, followed by staining with CellROX green reagent to detect ROS, according to the manufacturer's instructions (Thermo Fisher Scientific).

Cells were stained with the viable dye VIVID, followed by antibodies for human CD3 and CD8 (details as above), and anti-human CD45 APC-Cy7 (clone HI30, BioLegend) and anti-mouse/human CD11b PE-Cy7 (clone M1/70, BD Biosciences) as described previously [109].

Splenocytes were collected at 25 d following T cell transfer, then incubated with mouse and human FcR block (Miltenyi Biotec), stained with VIVID and antibodies for CD3, CD8, CD45, as above, and also with mouse anti-human pan HLA class I (clone W6/32, BioLegend).

Acknowledgements

A machine generated summary based on the work of Crowther, Michael D.; Dolton, Garry; Legut, Mateusz; Caillaud, Marine E.; Lloyd, Angharad; Attaf, Meriem; Galloway, Sarah A. E.; Rius, Cristina; Farrell, Colin P.; Szomolay, Barbara; Ager, Ann; Parker, Alan L.; Fuller, Anna; Donia, Marco; McCluskey, James; Rossjohn,

Jamie; Svane, Inge Marie; Phillips, John D.; Sewell, Andrew K. 2020 in Nature Immunology.

Lipid-droplet-accumulating microglia represent a dysfunctional and proinflammatory state in the aging brain.

https://doi.org/10.1038/s41593-019-0566-1

Abstract-Summary

We report a striking buildup of lipid droplets in microglia with aging in mouse and human brains.

These cells, which we call 'lipid-droplet-accumulating microglia' (LDAM), are defective in phagocytosis, produce high levels of reactive oxygen species and secrete proinflammatory cytokines.

An unbiased CRISPR–Cas9 screen identified genetic modifiers of lipid droplet formation; surprisingly, variants of several of these genes, including progranulin (GRN), are causes of autosomal-dominant forms of human neurodegenerative diseases.

Extended:

Targeting LDAM might represent an attractive and druggable approach to decrease neuroinflammation and to restore brain homeostasis in aging and neurodegeneration, with the goal of improving cognitive functions.

Main

Myeloid cells form lipid droplets in response to inflammation and stress, including the aforementioned macrophages in atherosclerotic lesions, leukocytes in inflammatory arthritis and eosinophils in allergic inflammation.

Lipid droplets have not been studied functionally in brain myeloid cells in humans or vertebrates, and less than a handful of papers report the histological presence of lipid droplets in human brains [110, 111].

Lipid droplets have been recognized for their role as inflammatory organelles in peripheral myeloid cells, and lipid droplets in glia have been rediscovered in the context of brain aging and disease.

Little is known about the formation and role of lipid droplets in microglia in vivo, and whether they have a role in neuroinflammation, brain aging or neurodegeneration.

We identify a novel state of microglia in the aging brain in which they accumulate lipid droplets.

Results

To explore whether LPS-induced lipid droplets in BV2 cells resemble those in microglia from aged mice, we compared their lipid composition using mass spectrometry.

These findings demonstrate that LPS triggers lipid droplet formation in microglia in vitro and in vivo, and that lipid-droplet-containing microglia in young LPS-treated mice show a transcriptional signature that partially overlaps with LDAM.

LPS-induced lipid droplets in BV2 cells are highly similar to lipid droplets in aged microglia, thus making this in vitro assay a useful model to study lipid droplets in LDAM.

We sought to explore the transcriptional commonalities of lipid-droplet-rich microglia in aging mice, in $Grn^{-/-}$ mice and in LPS-treated young mice.

LD-high microglia from $Grn^{-/-}$ mice and LDAM in aging mice have similar functional and transcriptional phenotypes, which suggests that lipid-droplet-containing $Grn^{-/-}$ microglia share the LDAM state.

Discussion

Microglia in the LDAM state, which account for more than 50% of all microglia in the aged hippocampus, but not microglia without lipid droplets, showed typical age-related functional impairments and a primed phenotype, which suggests that LDAM may be the primary detrimental microglia state in the aging brain.

We suggest that inflammation plays a key role in the buildup of lipid droplets in microglia, and it is tempting to speculate that age-related neuroinflammation provokes the formation of LDAM.

LDAM showed severe phagocytosis deficits compared with microglia without lipid droplets in the aging brain.

These lipid-droplet-rich microglia in $Grn^{-/-}$ mice showed a similar transcriptome signature and the same functional impairments as LDAM.

We showed that LDAM demonstrate a novel state of microglia with a unique transcriptional signature and functional impairments in the aging brain, and we identified lipid-droplet-containing microglia in a $Grn^{-/-}$ mouse model of chronic neuroinflammation and in an LPS-induced acutely inflamed brain milieu.

Methods

To assess ROS generation in primary microglia, cell homogenates from 3-month-old and 20-month-old male wild-type mice and from 9-month-old male $Grn^{-/-}$ mice were prepared, and antibody staining was performed as described above (see the section "Microglia isolation").

To measure ROS in BV2 cells, cells were split into 24-well plates at 5×10^4 cells per well in DMEM + 5% FBS and treated with LPS, triacsin C and vehicle solutions for 18 h. Next, cells were incubated in DMEM + 5%FBS with CellROX Orange (1:500; Invitrogen) for 30 min at 37 °C, washed twice in PBS, and CellROX Orange Intensity was examined by fluorescence microscopy (Keyence).

BV2 cells were treated for 18 h with 5 μg ml^{-1} LPS in DMEM with 5% FBS to induce lipid droplet formation, and the photoirradiation assay was performed as described above (see the section "iBP irradiation assay" above).

Acknowledgements

A machine generated summary based on the work of Marschallinger, Julia; Iram, Tal; Zardeneta, Macy; Lee, Song E.; Lehallier, Benoit; Haney, Michael S.; Pluvinage, John V.; Mathur, Vidhu; Hahn, Oliver; Morgens, David W.; Kim, Justin; Tevini, Julia; Felder, Thomas K.; Wolinski, Heimo; Bertozzi, Carolyn R.; Bassik, Michael C.; Aigner, Ludwig; Wyss-Coray, Tony 2020 in Nature Neuroscience.

Mitochondrial stress is relayed to the cytosol by an OMA1–DELE1–HRI pathway.

https://doi.org/10.1038/s41586-020-2078-2

Abstract-Summary

In mammalian cells, mitochondrial dysfunction triggers the integrated stress response, in which the phosphorylation of eukaryotic translation initiation factor 2α (eIF2α) results in the induction of the transcription factor ATF4 [112–114].

In a genome-wide CRISPR interference screen, we identified factors upstream of HRI: OMA1, a mitochondrial stress-activated protease; and DELE1, a little-characterized protein that we found was associated with the inner mitochondrial membrane.

Mitochondrial stress stimulates OMA1-dependent cleavage of DELE1 and leads to the accumulation of DELE1 in the cytosol, where it interacts with HRI and activates the eIF2α kinase activity of HRI.

The OMA1–DELE1–HRI pathway therefore represents a potential therapeutic target that could enable fine-tuning of the integrated stress response for beneficial outcomes in diseases that involve mitochondrial dysfunction.

Extended:

The OMA1–DELE1–HRI pathway we describe here could be targeted therapeutically to block the activation of ATF4 in cells that experience mitochondrial dysfunction, without globally blocking the ISR in all cells.

Main

The ISR is mediated by phosphorylation of eIF2α under various stress conditions, which are sensed by four eIF2α kinases [115, 116].

Phosphorylation of eIF2α reduces global protein synthesis but increases the translation of specific mRNAs—including ATF4, the master transcriptional regulator of the ISR.

HRI relays mitochondrial stress to ATF4

As the canonical mechanism of HRI activation is haem depletion [117], and key steps in haem biosynthesis occur in the mitochondria [118], we speculated that haem depletion could be the signal that activates HRI upon mitochondrial dysfunction.

CRISPRi screen reveals a role for DELE1

To express DELE1-mClover at lower levels, and enable its characterization without activating ATF4, we stably integrated it into the AAVS1 safe-harbour locus.

OMA1-dependent cleavage of DELE1

Because OMA1 and OPA1 localize to the inner mitochondrial membrane [119, 120], we investigated the localization of DELE1.

Mitochondrial dysfunction stimulates the OMA1-dependent cleavage of DELE1.

DELE1 interacts with and activates HRI

Our results suggest that the stress-induced accumulation of DELE1$_S$ in the cytosol leads to the activation of HRI through physical interaction.

In the context of endoplasmic-reticulum stress, there is a precedent for a factor (DDX3) being required in addition to eIF2α phosphorylation to induce ATF4 translation [121].

The failure to induce ATF4 in response to mitochondrial stress in DELE1-knockdown cells might result in pervasive cellular dysfunction, leading to the activation of other eIF2α kinases.

Role of the OMA1–DELE1–HRI pathway

Blockade of ATF4 induction led to upregulation, instead of downregulation, of cytosolic HSP70 (a molecular chaperone) under mitochondrial stress; this might contribute to the protective effect of HRI and DELE1 knockdown in some stress contexts.

Discussion

Both inhibition [122] and prolongation [123] of the ISR are potential therapeutic strategies.

The OMA1–DELE1–HRI pathway we describe here could be targeted therapeutically to block the activation of ATF4 in cells that experience mitochondrial dysfunction, without globally blocking the ISR in all cells.

Dissection of the molecular pathways that control different aspects of the ISR under different stress conditions might enable other therapeutic targets to be identified, which could enable fine-tuned manipulation of the cellular response to different stressors to achieve beneficial outcomes that are tailored to specific states of disease.

Methods

The protospacer sequences are: 5′-CGGGGTCCGCAAGCGCGAAG-3′ and 5′-AAACCCACTTCGTTCAAGAC-3′. The cell line expressing a DELE1-mClover transgene line from the AAVS1 locus was generated by transfecting a cXG289 population in which around 50% of cells expressed a DELE1 sgRNA and a BFP marker with pXG289 and TALENs that target the human AAVS1 locus (AAVS1-TALEN_L/R; Addgene 59025 and 59026).

Full-length N-terminal His$_6$-tagged rat HRI in construct pET28a-His$_6$-HRI [124] was co-expressed in Escherichia coli BL21(DE3) with chaperone plasmid pG-KJE8 (Takara Bio) at 15 °C for 72 h. Cleared cell lysates (lysed in Ni column buffer: 50 mM KPB pH 8.0, 300 mM NaCl, 10% glycerol, 5 mM imidazole and 0.5% CHAPS supplemented with protease inhibitor cocktail) were first applied to a His60 Ni Superflow column (Clontech).

Acknowledgements

A machine generated summary based on the work of Guo, Xiaoyan; Aviles, Giovanni; Liu, Yi; Tian, Ruilin; Unger, Bret A.; Lin, Yu-Hsiu T.; Wiita, Arun P.; Xu, Ke; Correia, M. Almira; Kampmann, Martin 2020 in Nature.

4.2   CRISPR Provides a Potential Disease Diagnosis and Treatment Strategy

Machine generated keywords: sample, liver, detection, mouse, correction, testing, patient, transmission, health, reaction, delivery, exon, adult, vivo, mrna.

A dual AAV system enables the Cas9-mediated correction of a metabolic liver disease in newborn mice.

https://doi.org/10.1038/nbt.3469

Main

We developed a strategy using a dual-AAV system based on AAV8, which has high liver tropism, to correct the point mutation in newborn spf^{ash} mice using Cas9 enzyme from Staphylococcus aureus (SaCas9) [125–127].

Vector 1 expressed the SaCas9 gene from a liver-specific TBG promoter (subsequently referred to as AAV8.SaCas9), whereas vector 2 contained the sgRNA1 sequence expressed from the U6 promoter and the 1.8-kb donor OTC DNA sequence (referred to as AAV8.sgRNA1.donor).

We obtained liver samples from treated spf^{ash} animals, untreated spf^{ash} (spf^{ash} controls), wild-type littermates and spf^{ash} mice administered AAV8.SaCas9 with a modified AAV8.control.donor without guide RNA (untargeted) at 1, 3 and 8 weeks following vector infusion.

Methods

To generate a dual AAV vector system for in vivo OTC gene correction by SaCas9, we constructed two AAV cis-plasmids: (i) the hSaCas9 was subcloned from pX330.hSaCas9 into an AAV backbone plasmid containing the full-length TBG promoter (two copies of enhancer elements of the α microglobulin/bikunin gene followed by a liver-specific TBG promoter) and the bovine growth hormone polyadenylation sequence, yielding AAV8.SaCas9; (ii) the 1.8-kb OTC donor template was cloned into the pAAV backbone, and the U6-OTC sgRNA1 cassette was inserted into the AflII site, yielding AAV8.sgRNA1.donor.

The following assays were performed in a blinded fashion in which the investigator was unaware of the nature of the vectors or vector dose: vector injection, OTC and Cas9 (FLAG) immunostaining and quantification, histopathology analysis on liver, OTC enzyme activity assay, and gene expression analysis and RT-qPCR.

Acknowledgements

A machine generated summary based on the work of Yang, Yang; Wang, Lili; Bell, Peter; McMenamin, Deirdre; He, Zhenning; White, John; Yu, Hongwei; Xu, Chenyu; Morizono, Hiroki; Musunuru, Kiran; Batshaw, Mark L; Wilson, James M 2016 in Nature Biotechnology.

Correction of a pathogenic gene mutation in human embryos.

https://doi.org/10.1038/nature23305

Abstract-Summary

We describe the correction of the heterozygous MYBPC3 mutation in human preimplantation embryos with precise CRISPR–Cas9-based targeting accuracy and high homology-directed repair efficiency by activating an endogenous, germline-specific DNA repair response.

By modulating the cell cycle stage at which the DSB was induced, we were able to avoid mosaicism in cleaving embryos and achieve a high yield of homozygous embryos carrying the wild-type MYBPC3 gene without evidence of off-target mutations.

The efficiency, accuracy and safety of the approach presented suggest that it has potential to be used for the correction of heritable mutations in human embryos by complementing preimplantation genetic diagnosis.

Main

MYBPC3 mutations account for approximately 40% of all genetic defects causing HCM and are also responsible for a large fraction of other inherited cardiomyopathies, including dilated cardiomyopathy and left ventricular non-compaction [128].

In an effort to demonstrate the proof-of-principle that heterozygous gene mutations can be corrected in human gametes or early embryos, we focused on the MYBPC3 mutation that has been implicated in HCM.

Although homozygous mutations with no PGD alternative would have been most desirable for gene correction, generating homozygous human embryos for research purposes is practically impossible.

We specifically targeted the heterozygous four-base-pair (bp) deletion in the MYBPC3 gene in human zygotes introduced by heterozygous, carrier sperm while oocytes obtained from healthy donors provided the wild-type allele.

DSBs in the mutant paternal MYBPC3 gene were preferentially repaired using the wild-type oocyte allele as a template, suggesting an alternative, germline-specific DNA repair response.

[Section 2]

Of 61 iPSC clones transfected with CRISPR–Cas9-1, 44 (72.1%) were not targeted, as evidenced by the presence of both intact wild-type and intact mutant alleles.

The targeting efficiency with CRISPR–Cas9-2 was 13.1% (23/175) and the HDR was considerably lower at 13% (3/23).

The wild-type allele in all iPSC clones analysed remained intact, demonstrating high fidelity of sgRNAs.

On-target mutations were not detected in wild-type embryonic stem (ES) cells (H9) carrying both wild-type MYBPC3 alleles, demonstrating high specificity of CRISPR–Cas9-1.

[Section 3]

As these embryos originated from $MYBPC3^{WT/\Delta GAGT}$ zygotes, individual blastomeres are likely to have repaired the $MYBPC3^{\Delta GAGT}$ deletion by HDR using the maternal wild-type allele as a template instead of the injected ssODNs.

We did not find any evidence of HDR using exogenous ssODN, suggesting that HDR is guided exclusively by the wild-type maternal allele.

As HDR calculations are based on mosaic embryos only, it is probable that some targeted heterozygous ($MYBPC3^{WT/\Delta GAGT}$) zygotes repaired the mutant allele in all blastomeres using the wild-type template ($MYBPC3^{WT/WT}$).

These HDR-repaired, uniform embryos would be indistinguishable from their wild-type homozygous counterparts, thereby increasing the portion of $MYBPC3^{WT/WT}$ embryos in the CRISPR–Cas9 injected group.

Targeting gametes eliminates mosaicism

There are two different possible explanations for this outcome: (1) at the time of injection, a zygote had completed the S-phase of the cell cycle with DNA replication and already produced two mutant alleles [129]; (2) CRISPR–Cas9 remained active, continuing to target after zygotic division.

At day 3 after fertilization, embryos were disaggregated and each individual blastomere was analysed as described above for S-phase-injected zygotes.

All sister blastomeres in all but one embryo carried identical genotypes, indicating a marked reduction in mosaicism in M-phase-injected embryos.

This embryo, with every blastomere carrying repaired $MYBPC3^{WT/WT}$, would be eligible for transfer.

Development and cytogenetics of repaired embryos

In an effort to provide additional insights into the developmental competence of gene-corrected blastocysts, and to obtain sufficient cellular material for detailed cytogenetic studies, we established six ES cell lines from CRISPR–Cas9-injected blastocysts and one from controls.

CRISPR–Cas9-treated human embryos displayed normal development to blasto-cysts and ES cells without cytogenetic abnormalities.

Off-target consequences in repaired human embryos

We conducted a comprehensive, whole-genome sequencing (WGS) analysis of the patient's genomic DNA by digested genome sequencing (Digenome-seq) [130, 131].

Potential off-target sequences were identified by digestion of iPSC-derived, cell-free genomic DNA with CRISPR–Cas9 followed by WGS.

Sequencing reads of CRISPR–Cas9-digested genomic DNA are vertically aligned at on-/off-target sites in IGV viewer [130, 132].

Improved Digenome-seq provides DNA cleavage scores for potential off-target sites based on alignment patterns of sequence reads [131].

We also investigated whether CRISPR–Cas9 targeting induced global off-target genetic variations and genome instability by performing whole-exome sequencing (WES) in CRISPR–Cas9-treated ES cells and compared the results to those of control ES cells and corresponding egg and sperm donor blood DNA.

Discussion

We show here that DSBs in human gametes and zygotes are preferentially resolved using an endogenous HDR mechanism, exclusively directed by the wild-type allele as a repair template.

In a study involving human heterozygous embryos, HDR was exclusively directed by the exogenous DNA template with no evidence of wild-type allele-based repair [133].

Similar to the outcomes seen in mice and other animals, genome-edited human preimplantation embryos and newborn monkeys display mosaic targeting genotypes in their cells and tissues, suggesting that DSBs and subsequent repair do not occur at the single mutant allele stage [133–135].

We demonstrate that the delivery of CRISPR–Cas9 into M-phase oocytes abolished mosaicism in cleaving embryos, suggesting that gene targeting and editing efficiencies are strongly associated with DNA synthesis and the cell cycle phase [136].

It may be challenging to repair homozygous mutations in human embryos when both alleles are mutant and wild-type allele-based HDR mechanisms cannot be employed.

Methods

The regulatory framework surrounding the use of human gametes and embryos for this research was based on the guidelines set by the Oregon Health & Science University (OHSU) Stem Cell Research Oversight Committee (OSCRO).

The established track record of the study team to uphold strict confidentiality and regulatory requirements paved the way for full OHSU IRB study approval in 2016, contingent upon strict continuing oversight which includes: a phased scientific approach requiring evaluation of results on the safety and efficacy of germline gene correction in iPSCs before approving studies on human pre-implantation embryos; external bi-annual monitoring of all regulatory documents regarding human subjects; bi-annual Data Safety Monitoring Committee (DSMC) review; and annual continuing review by the OHSU IRB.

Acknowledgements

A machine generated summary based on the work of Ma, Hong; Marti-Gutierrez, Nuria; Park, Sang-Wook; Wu, Jun; Lee, Yeonmi; Suzuki, Keiichiro; Koski, Amy; Ji, Dongmei; Hayama, Tomonari; Ahmed, Riffat; Darby, Hayley; Van Dyken, Crystal; Li, Ying; Kang, Eunju; Park, A.-Reum; Kim, Daesik; Kim, Sang-Tae; Gong, Jianhui; Gu, Ying; Xu, Xun; Battaglia, David; Krieg, Sacha A.; Lee, David M.; Wu, Diana H.; Wolf, Don P.; Heitner, Stephen B.; Belmonte, Juan Carlos Izpisua; Amato, Paula; Kim, Jin-Soo; Kaul, Sanjiv; Mitalipov, Shoukhrat 2017 in Nature.

Development of a gene-editing approach to restore vision loss in Leber congenital amaurosis type 10.

https://doi.org/10.1038/s41591-018-0327-9

Abstract-Summary

We developed EDIT-101, a candidate genome-editing therapeutic, to remove the aberrant splice donor created by the IVS26 mutation in the CEP290 gene and restore normal CEP290 expression.

Key to this therapeutic, we identified a pair of Staphylococcus aureus Cas9 guide RNAs that were highly active and specific to the human CEP290 target sequence.

Subretinal delivery of EDIT-101 in humanized CEP290 mice showed rapid and sustained CEP290 gene editing.

A comparable surrogate non-human primate (NHP) vector also achieved productive editing of the NHP CEP290 gene at levels that met the target therapeutic threshold, and demonstrated the ability of CRISPR/Cas9 to edit somatic primate cells in vivo.

Main

Subretinal delivery of EDIT-101, or an NHP surrogate vector, was well tolerated and achieved sustained and dose-dependent CEP290 editing in photoreceptor cells, in mice and non-human primates, that met or exceeded the target therapeutic level.

These results demonstrate that targeted gene editing of the IVS26 mutation-containing region of CEP290 restores correct splicing and expression of CEP290, validating the molecular mechanism of our gene-editing strategy.

We used a human CEP290 IVS26 knock-in mouse model [137] to assess the kinetics and dose response of targeted gene-editing efficiency in vivo following the subretinal delivery of EDIT-101.

The data from both species demonstrate productive editing rates in >10% of photoreceptors, with doses ranging from 3×10^{11} to 3×10^{12} vg ml^{-1}, and this aligns well with the 5×10^{11} vg ml^{-1} dose of voretigene neparvovec-rzyl (Luxturna, Spark Therapeutics) [138].

Methods

Cells were plated in a non-tissue culture-treated T-75 flask at a concentration of 1.3 $\times 10^6$ cells ml^{-1} and incubated at 37 °C for 48 h. After 48 h, beads were removed from cells by magnetic separation.

In one well of a 96-well plate, 5 µl of 70 ng µl^{-1} genomic DNA prepared as described was mixed with 5 µl of RNP dilution or × 1 H150 buffer with 2 mM MgCl^2, creating a three-log dose response ranging from 1.0 to 0.01 µM for each RNP of interest and an untreated control.

Sequencing libraries were created as previously described by Tsai and others [139] with the following DNA shearing modification: 400 ng of genomic was sheared in 130 µl × 1 TE Buffer (Tris–EDTA: 10 mM Tris base, 1 mM EDTA, pH 8.0) with adaptive focused acoustics (Covaris M220, Covaris, Inc.) (peak power, 50 W; cycles/burst, 200; duration, 60 s; duty factor, 10) to obtain 500 bp fragments.

Acknowledgements

A machine generated summary based on the work of Maeder, Morgan L.; Stefanidakis, Michael; Wilson, Christopher J.; Baral, Reshica; Barrera, Luis Alberto; Bounoutas, George S.; Bumcrot, David; Chao, Hoson; Ciulla, Dawn M.; DaSilva, Jennifer A.; Dass, Abhishek; Dhanapal, Vidya; Fennell, Tim J.; Friedland, Ari E.; Giannoukos, Georgia; Gloskowski, Sebastian W.; Glucksmann, Alexandra; Gotta, Gregory M.; Jayaram, Hariharan; Haskett, Scott J.; Hopkins, Bei; Horng, Joy E.; Joshi, Shivangi; Marco, Eugenio; Mepani, Rina; Reyon, Deepak; Ta, Terence; Tabbaa, Diana G.; Samuelsson, Steven J.; Shen, Shen; Skor, Maxwell N.; Stetkiewicz, Pam; Wang, Tongyao; Yudkoff, Clifford; Myer, Vic E.; Albright, Charles F.; Jiang, Haiyan 2019 in Nature Medicine.

CRISPR–Cas12-based detection of SARS-CoV-2.

https://doi.org/10.1038/s41587-020-0513-4

Abstract-Summary

COVID-19, the disease associated with SARS-CoV-2 infection, rapidly spread to produce a global pandemic.

We report development of a rapid (<40 min), easy-to-implement and accurate CRISPR–Cas12-based lateral flow assay for detection of SARS-CoV-2 from respiratory swab RNA extracts.

Our CRISPR-based DETECTR assay provides a visual and faster alternative to the US Centers for Disease Control and Prevention SARS-CoV-2 real-time RT–PCR assay, with 95% positive predictive agreement and 100% negative predictive agreement.

Main

To accelerate clinical diagnostic testing for COVID-19 in the United States, on 28 February 2020, the US Food and Drug Administration (FDA) permitted individual clinically licensed laboratories to report the results of in-house-developed SARS-CoV-2 diagnostic assays while awaiting results of an EUA submission for approval [140].

We report the development and initial validation of a CRISPR–Cas12-based assay [141–145] for detection of SARS-CoV-2 from extracted patient sample RNA, called SARS-CoV-2 DNA Endonuclease-Targeted CRISPR Trans Reporter (DETECTR).

The positive predictive agreement and negative predictive agreement of SARS-CoV-2 DETECTR relative to the CDC qRT–PCR assay were 95% and 100%, respectively, for detection of the coronavirus in 83 total respiratory swab samples.

We combined isothermal amplification with CRISPR–Cas12 DETECTR technology to develop a rapid (30–40 min) test for detection of SARS-CoV-2 in clinical samples.

Methods

DETECTR assays were performed using RT–LAMP for preamplification of viral or control RNA targets and LbCas12a for the trans-cleavage assay.

The patient-optimized DETECTR assays were performed using the RT–LAMP method as described above with the following modifications: a DNA-binding dye, SYTO9 (Thermo Fisher Scientific) was included in the reaction to monitor the amplification reaction and the incubation time was extended to 30 min to capture data from lower titer samples.

The fluorescence-based patient-optimized LbCas12a trans-cleavage assays were performed as described above with modifications; 40 nM LbCas12a was preincubated with 40 nM gRNA, after which 100 nM of a fluorescent reporter molecule compatible with detection in the presence of the SYTO9 dye (/5Alex594N/TTATTATT/3IAbRQSp/) was added to the complex.

Sample RNA of SARS-CoV-2 was extracted following instructions as described in the CDC EUA-approved protocol [146] (input 120 µl, elution of 120 µl) using Qiagen DSP Viral RNA Mini kit (Qiagen) at CDPH and the MagNA Pure 24 instrument (Roche Life Science) at UCSF.

Acknowledgements

A machine generated summary based on the work of Broughton, James P.; Deng, Xianding; Yu, Guixia; Fasching, Clare L.; Servellita, Venice; Singh, Jasmeet; Miao, Xin; Streithorst, Jessica A.; Granados, Andrea; Sotomayor-Gonzalez, Alicia; Zorn, Kelsey; Gopez, Allan; Hsu, Elaine; Gu, Wei; Miller, Steve; Pan, Chao-Yang; Guevara, Hugo; Wadford, Debra A.; Chen, Janice S.; Chiu, Charles Y. 2020 in Nature Biotechnology.

Massively multiplexed nucleic acid detection with Cas13.

https://doi.org/10.1038/s41586-020-2279-8

Abstract-Summary

To enable routine surveillance and comprehensive diagnostic applications, there is a need for detection technologies that can scale to test many samples [147–149] while simultaneously testing for many pathogens [150–152].

We develop Combinatorial Arrayed Reactions for Multiplexed Evaluation of Nucleic acids (CARMEN), a platform for scalable, multiplexed pathogen detection.

In the CARMEN platform, nanolitre droplets containing CRISPR-based nucleic acid detection reagents [143] self-organize in a microwell array [153] to pair with droplets of amplified samples, testing each sample against each CRISPR RNA (crRNA) in replicate.

The combination of CARMEN and Cas13 detection (CARMEN–Cas13) enables robust testing of more than 4500 crRNA–target pairs on a single array.

Scalable, highly multiplexed CRISPR-based nucleic acid detection shifts diagnostic and surveillance efforts from targeted testing of high-priority samples to comprehensive testing of large sample sets, greatly benefiting patients and public health [154–156].

Main

We envisioned that CRISPR-based nucleic acid detection could be integrated with the microwell-array system to test many amplified samples for many analytes in parallel.

Accurate testing of multiple samples for hundreds of microbial pathogens requires higher throughput than is offered by existing multiplexed detection systems [148, 157, 158].

To enable highly multiplexed detection with high sample throughput, we developed a set of 1050 solution-based colour codes using ratios of 4 commercially available, small-molecule fluorophores.

We have demonstrated a broad set of uses for CARMEN–Cas13 in differentiating viral sequences at the species, strain, and single-nucleotide polymorphism (SNP) levels and the capability to rapidly develop and validate highly multiplexed detection panels.

We imagine region- and outbreak-specific detection panels deployed to test thousands of samples from selected populations, including animal vectors, animal reservoirs, or patients presenting with symptoms.

Methods

The chip was imaged with a 1× objective to identify colour codes, droplet pairs were merged, and reporter fluorescence in each well was measured by fluorescence imaging at 1 h and 3 h (see 'Loading, imaging and merging microwell arrays' under 'General procedures').

The chip was imaged with a 2× objective to identify colour codes, droplet pairs were merged, and reporter fluorescence in each well was measured by fluorescence imaging at 1 or 3 h (see 'Loading, imaging and merging microwell arrays' under 'General procedures').

The chip was imaged with a 2× objective to identify colour codes, droplet pairs were merged, and reporter fluorescence in each well was measured by fluorescence imaging at 30 min or 3 h (see 'Loading, imaging and merging microwell arrays' under 'General procedures').

Acknowledgements

A machine generated summary based on the work of Ackerman, Cheri M.; Myhrvold, Cameron; Thakku, Sri Gowtham; Freije, Catherine A.; Metsky, Hayden C.; Yang, David K.; Ye, Simon H.; Boehm, Chloe K.; Kosoko-Thoroddsen, Tinna-Sólveig F.; Kehe, Jared; Nguyen, Tien G.; Carter, Amber; Kulesa, Anthony; Barnes, John R.; Dugan, Vivien G.; Hung, Deborah T.; Blainey, Paul C.; Sabeti, Pardis C. 2020 in Nature.

In vivo *adenine base editing of PCSK9 in macaques reduces LDL cholesterol levels.*

https://doi.org/10.1038/s41587-021-00933-4

Abstract-Summary

We investigated the efficacy and safety of ABEs in the livers of mice and cynomolgus macaques for the reduction of blood low-density lipoprotein (LDL) levels.

Lipid nanoparticle–based delivery of mRNA encoding an ABE and a single-guide RNA targeting PCSK9, a negative regulator of LDL, induced up to 67% editing (on average, 61%) in mice and up to 34% editing (on average, 26%) in macaques.

Plasma PCSK9 and LDL levels were stably reduced by 95% and 58% in mice and by 32% and 14% in macaques, respectively.

Re-dosing in macaques did not increase editing, possibly owing to the detected humoral immune response to ABE upon treatment.

Extended:

We applied base editing to install a splice site mutation in PCSK9, and we show here that LNP-mediated delivery of ABE-encoding nucleoside-modified mRNA, together with chemically modified sgRNA, results in editing rates of up to 30% in the liver of macaques.

Main

Programmable CRISPR–Cas nucleases enable genome editing by generating double-stranded DNA breaks at the target locus [159].

They convert C•T into T•A or A•T into G•C base pairs without the requirement of homology-directed repair and, therefore, enable precise and efficient editing in tissues with slow turnover rates, such as the liver [160–163].

Considering that most pathogenic point mutations are C•G to T•A conversions, ABEs are of particular interest for in vivo genome editing therapies [161].

For clinical application of base editing, the potential generation of off-target mutations represents a major concern.

We report that lipid nanoparticle (LNP)-mediated delivery of ABE-encoding nucleoside-modified mRNA, together with a chemically stabilized sgRNA, enables efficient editing in mice and NHPs without inducing off-target mutations on genomic DNA.

Results

Recent ex vivo studies in human induced pluripotent stem cells, two-cell stage mouse embryos and rice callus cells associated cytidine base editor (CBE) expression, but not ABE expression, with sgRNA-independent off-target deamination in genomic DNA [164–166].

To assess whether this observation holds true for in vivo adenine base editing in the liver, we analyzed the genomic DNA of hepatocytes from AAV- and LNP-treated animals by whole-genome sequencing (WGS).

We found no evidence for the generation of substantial off-target mutations in genomic DNA after AAV- or LNP-mediated adenine base editing in vivo.

To assess hepatotropism of the applied LNP formulation, we analyzed on-target editing rates in nine different organs from animals of the high-dose groups.

The top candidate off-target sites in the human and macaque genome were then analyzed by deep sequencing in human HepG2 cells transfected with plasmids expressing ABEmax and sgRNA_hP01 and in macaques of the high-dose groups.

Discussion

We applied base editing to install a splice site mutation in PCSK9, and we show here that LNP-mediated delivery of ABE-encoding nucleoside-modified mRNA, together with chemically modified sgRNA, results in editing rates of up to 30% in the liver of macaques.

One of the major considerations for the clinical development of safe and effective genome editing therapies is the minimization or elimination of off-target editing and mutations.

We demonstrate that transient LNP-mediated ABE delivery into mouse livers enables up to 88% on-target editing in hepatocytes without inducing substantial sgRNA-dependent or sgRNA-independent (unguided) off-target editing in genomic DNA.

We provide preclinical data in mice and NHPs for the application of adenine base editing to treat genetic liver diseases.

We demonstrate therapeutically beneficial editing using transient and non-viral delivery vectors without induction of considerable off-target mutations.

Methods

Genomic DNA from mouse tissues was isolated using the DNeasy Blood and Tissue Kit (Qiagen) according to the manufacturer's protocol or directly lysed using direct lysis buffer: 10 µl of 4 × lysis buffer (10 mM Tris–HCl pH 8, 2% Triton X-100, 1 mM EDTA, 1% freshly added proteinase K) and incubated at 60 °C for 60 min and 95 °C for 10 min.

We considered variants at autosomal chromosomes without any evidence from a paired control sample (genomic DNA isolated from untreated tissue from the same mouse); passed by VariantFiltration with a GATK phred-scaled quality score ≥ 100 for base substitutions and ≥ 250 for indels; a base coverage of at least $20\times$ in the clonal and paired control sample; mapping quality ≥ 60; and no overlap with single-nucleotide polymorphisms in the Single Nucleotide Polymorphism Database version 142.

Acknowledgements

A machine generated summary based on the work of Rothgangl, Tanja; Dennis, Melissa K.; Lin, Paulo J. C.; Oka, Rurika; Witzigmann, Dominik; Villiger, Lukas; Qi, Weihong; Hruzova, Martina; Kissling, Lucas; Lenggenhager, Daniela; Borrelli, Costanza; Egli, Sabina; Frey, Nina; Bakker, Noëlle; Walker, John A.; Kadina, Anastasia P.; Victorov, Denis V.; Pacesa, Martin; Kreutzer, Susanne; Kontarakis, Zacharias; Moor, Andreas; Jinek, Martin; Weissman, Drew; Stoffel, Markus; van Boxtel, Ruben; Holden, Kevin; Pardi, Norbert; Thöny, Beat; Häberle, Johannes; Tam, Ying K.; Semple, Sean C.; Schwank, Gerald 2021 in Nature Biotechnology.

References

1. Vermeulen, L., et al. 2013. Defining stem cell dynamics in models of intestinal tumor initiation. *Science* 342: 995–998.
2. Fearon, E.R. 2011. Molecular genetics of colorectal cancer. *Annual Review of Pathology: Mechanisms of Disease* 6: 479–507.
3. Barker, N., et al. 2009. Crypt stem cells as the cells-of-origin of intestinal cancer. *Nature* 457: 608–611.
4. Schepers, A.G., et al. 2012. Lineage tracing reveals Lgr5+ stem cell activity in mouse intestinal adenomas. *Science* 337: 730–735.
5. Zhu, L., et al. 2009. Prominin 1 marks intestinal stem cells that are susceptible to neoplastic transformation. *Nature* 457: 603–607.
6. Li, X., et al. 2014. Oncogenic transformation of diverse gastrointestinal tissues in primary organoid culture. *Nature Medicine* 20: 769–777.
7. Mali, P., et al. 2013. RNA-guided human genome engineering via Cas9. *Science* 339: 823–826.
8. Cho, S.W., S. Kim, J.M. Kim, and J.S. Kim. 2013. Targeted genome engineering in human cells with the Cas9 RNA-guided endonuclease. *Nature Biotechnology* 31: 230–232.
9. Cong, L., et al. 2013. Multiplex genome engineering using CRISPR/Cas systems. *Science* 339: 819–823.
10. Shi, Y., A. Hata, R.S. Lo, J. Massaqué, and N.P. Pavletich. 1997. A structural basis for mutational inactivation of the tumour suppressor Smad4. *Nature* 388: 87–93.
11. Chacko, B.M., et al. 2004. Structural basis of heteromeric smad protein assembly in TGF-β signaling. *Molecular Cell* 15: 813–823.
12. Huch, M., et al. 2013. In vitro expansion of single Lgr5+ liver stem cells induced by Wnt-driven regeneration. *Nature* 494: 247–250.
13. Brentjens, R.J. et al. 2013. CD19-targeted T cells rapidly induce molecular remissions in adults with chemotherapy-refractory acute lymphoblastic leukemia. Science Translational Medicine 5: 177ra38.

14. Brentjens, R.J., et al. 2003. Eradication of systemic B-cell tumors by genetically targeted human T lymphocytes co-stimulated by CD80 and interleukin-15. *Nature Medicine* 9: 279–286.
15. Zhao, Z., et al. 2015. Structural design of engineered costimulation determines tumor rejection kinetics and persistence of CAR T cells. *Cancer Cell* 28: 415–428.
16. Maude, S.L., et al. 2014. Chimeric antigen receptor T cells for sustained remissions in leukemia. *New England Journal of Medicine* 371: 1507–1517.
17. Shalem, O., et al. 2014. Genome-scale CRISPR-Cas9 knockout screening in human cells. *Science* 343: 84–87.
18. Sáez-Borderías, A., et al. 2009. IL-12-dependent inducible expression of the CD94/NKG2A inhibitory receptor regulates CD94/NKG2C+ NK cell function. *The Journal of Immunology* 182: 829–836.
19. Moser, J.M., J. Gibbs, P.E. Jensen, and A.E. Lukacher. 2002. CD94-NKG2A receptors regulate antiviral CD8+ T cell responses. *Nature Immunology* 3: 189–195.
20. Hu, D., et al. 2004. Analysis of regulatory CD8 T cells in Qa-1-deficient mice. *Nature Immunology* 5: 516–523.
21. Wang, T., et al. 2015. Identification and characterization of essential genes in the human genome. *Science* 350: 1096–1101.
22. Aguirre, A.J., et al. 2016. Genomic copy number dictates a gene-independent cell response to CRISPR/Cas9 targeting. *Cancer Discovery* 6: 914–929.
23. Munoz, D.M., et al. 2016. CRISPR screens provide a comprehensive assessment of cancer vulnerabilities but generate false-positive hits for highly amplified genomic regions. *Cancer Discovery* 6: 900–913.
24. Tsherniak, A., et al. 2017. Defining a cancer dependency map. *Cell* 170: 564-576.e16.
25. Morgens, D.W., et al. 2017. Genome-scale measurement of off-target activity using Cas9 toxicity in high-throughput screens. *Nature Communications* 8: 15178.
26. Wilson, W.H., et al. 2015. Targeting B cell receptor signaling with ibrutinib in diffuse large B cell lymphoma. *Nature Medicine* 21: 922–926.
27. Young, R.M., et al. 2015. Survival of human lymphoma cells requires B-cell receptor engagement by self-antigens. *Proceedings of the National Academy of Sciences of the United States of America* 112: 13447–13454.
28. Davis, R.E., et al. 2010. Chronic active B-cell-receptor signalling in diffuse large B-cell lymphoma. *Nature* 463: 88–92.
29. Lenz, G., et al. 2008. Oncogenic CARD11 mutations in human diffuse large B cell lymphoma. *Science* 319: 1676–1679.
30. Ngo, V.N., et al. 2011. Oncogenically active MYD88 mutations in human lymphoma. *Nature* 470: 115–119.
31. Panier, S., and S.J. Boulton. 2014. Double-strand break repair: 53BP1 comes into focus. *Nature Reviews Molecular Cell Biology* 15: 7–18.
32. Hustedt, N., and D. Durocher. 2016. *Nature Cell Biology* 19: 1–9.
33. Fradet-Turcotte, A., et al. 2013. 53BP1 is a reader of the DNA-damage-induced H2A Lys15 ubiquitin mark. *Nature* 499: 50–54.
34. Adkins, N.L., H. Niu, P. Sung, and C.L. Peterson. 2013. Nucleosome dynamics regulates DNA processing. *Nature Structural & Molecular Biology* 20: 836–842.
35. Jaspers, J.E., et al. 2013. Loss of 53BP1 causes PARP inhibitor resistance in Brca1-mutated mouse mammary tumors. *Cancer Discovery* 3: 68–81.
36. Bouwman, P., et al. 2010. 53BP1 loss rescues BRCA1 deficiency and is associated with triple-negative and BRCA-mutated breast cancers. *Nature Structural & Molecular Biology* 17: 688–695.
37. Bunting, S.F., et al. 2010. 53BP1 inhibits homologous recombination in Brca1-deficient cells by blocking resection of DNA breaks. *Cell* 141: 243–254.
38. Carapetis, J.R., A.C. Steer, E.K. Mulholland, and M. Weber. 2005. The global burden of group A streptococcal diseases. *The Lancet Infectious Diseases* 5: 685–694.

39. Charlesworth, C.T. et al. 2018. Identification of pre-existing adaptive immunity to Cas9 proteins in humans. Preprint at https://www.biorxiv.org/content/early/2018/01/05/243345.
40. Kawalekar, O.U., R.S. O'Connor, J.A. Fraietta, L. Guo, S.E. McGettigan, A.D. Posey Jr., P.R. Patel, S. Guedan, J. Scholler, B. Keith, N.W. Snyder, I.A. Blair, M.C. Milone, and C.H. June. 2016. Distinct signaling of coreceptors regulates specific metabolism pathways and impacts memory development in CAR T cells. *Immunity* 44 (2): 380–390.
41. Moon, E.K., L.C. Wang, D.V. Dolfi, C.B. Wilson, R. Ranganathan, J. Sun, V. Kapoor, J. Scholler, E. Pure, M.C. Milone, C.H. June, J.L. Riley, E.J. Wherry, and S.M. Albelda. 2014. Multifactorial T-cell hypofunction that is reversible can limit the efficacy of chimeric Antigen receptor-transduced human T cells in solid tumors. *Clinical Cancer Research* 20 (16): 4262–4273.
42. Tchou, J., L.C. Wang, B. Selven, H. Zhang, J. Conejo-Garcia, H. Borghaei, M. Kalos, R.H. Vondeheide, S.M. Albelda, C.H. June, and P.J. Zhang. 2012. Mesothelin, a novel immunotherapy target for triple negative breast cancer. *Breast Cancer Research and Treatment* 133 (2): 799–804.
43. Ren, J., X. Liu, C. Fang, S. Jiang, C.H. June, and Y. Zhao. 2017. Multiplex genome editing to generate universal CAR T cells resistant to PD1 inhibition. *Clinical Cancer Research* 23 (9): 2255–2266.
44. Rupp, L.J., K. Schumann, K.T. Roybal, R.E. Gate, C.J. Ye, W.A. Lim, and A. Marson. 2017. CRISPR/Cas9-mediated PD-1 disruption enhances anti-tumor efficacy of human chimeric antigen receptor T cells. *Science and Reports* 7 (1): 737.
45. Su, S., Z. Zou, F. Chen, N. Ding, J. Du, J. Shao, L. Li, Y. Fu, B. Hu, Y. Yang, H. Sha, F. Meng, J. Wei, X. Huang, and B. Liu. 2017. CRISPR-Cas9-mediated disruption of PD-1 on human T cells for adoptive cellular therapies of EBV positive gastric cancer. *Oncoimmunology* 6(1): e1249558.
46. Ren, J., X. Zhang, X. Liu, C. Fang, S. Jiang, C.H. June, and Y. Zhao. 2017. A versatile system for rapid multiplex genome-edited CAR T cell generation. *Oncotarget* 8 (10): 17002–17011.
47. Cherkassky, L., A. Morello, J. Villena-Vargas, Y. Feng, D.S. Dimitrov, D.R. Jones, M. Sadelain, and P.S. Adusumilli. 2016. Human CAR T cells with cell-intrinsic PD-1 checkpoint blockade resist tumor-mediated inhibition. *The Journal of Clinical Investigation* 126 (8): 3130–3144.
48. Lowy, F.D. 1998. Staphylococcus aureus infections. *New England Journal of Medicine* 339: 520–532.
49. Roberts, A.L., et al. 2012. Detection of group A Streptococcus in tonsils from pediatric patients reveals high rate of asymptomatic streptococcal carriage. *BMC Pediatrics* 12: 3.
50. Czerkinsky, C.C., L.A. Nilsson, H. Nygren, O. Ouchterlony, and A. Tarkowski. 1983. A solid-phase enzyme-linked immunospot (ELISPOT) assay for enumeration of specific antibody-secreting cells. *Journal of Immunological Methods* 65: 109–121.
51. Lovelace, P., and H.T. Maecker. 2011. Multiparameter intracellular cytokine staining. *Methods in Molecular Biology* 699: 165–178.
52. Frentsch, M., et al. 2005. Direct access to CD4+ T cells specific for defined antigens according to CD154 expression. *Nature Medicine* 11: 1118–1124.
53. Bacher, P., et al. 2013. Antigen-reactive T cell enrichment for direct, high-resolution analysis of the human naive and memory Th cell repertoire. *The Journal of Immunology* 190: 3967–3976.
54. Garraway, L.A. 2013. Genomics-driven oncology: Framework for an emerging paradigm. *Journal of Clinical Oncology* 31: 1806–1814.
55. Hart, T., et al. 2015. High-resolution CRISPR screens reveal fitness genes and genotype-specific cancer liabilities. *Cell* 163: 1515–1526.
56. Hart, T. et al. 2017. Evaluation and design of genome-wide CRISPR/SpCas9 knockout screens. G3 (Bethesda) 7, 2719–2727.
57. Iorio, F., et al. 2016. A landscape of pharmacogenomic interactions in cancer. *Cell* 166: 740–754.

58. Chu, W.K., and I.D. Hickson. 2009. RecQ helicases: Multifunctional genome caretakers. *Nature Reviews Cancer* 9: 644–654.

59. Perry, J.J.P., et al. 2006. WRN exonuclease structure and molecular mechanism imply an editing role in DNA end processing. *Nature Structural & Molecular Biology* 13: 414–422.

60. Kamath-Loeb, A.S., P. Welcsh, M. Waite, E.T. Adman, and L.A. Loeb. 2004. The enzymatic activities of the Werner syndrome protein are disabled by the amino acid polymorphism R834C. *Journal of Biological Chemistry* 279: 55499–55505.

61. Ketkar, A., M. Voehler, T. Mukiza, and R.L. Eoff. 2017. Residues in the RecQ C-terminal domain of the human Werner Syndrome helicase are involved in unwinding G-quadruplex DNA. *Journal of Biological Chemistry* 292: 3154–3163.

62. Sillman, B., et al. 2018. Creation of a long-acting nanoformulated dolutegravir. *Nature Communications* 9: 443.

63. Dash, P.K., et al. 2012. Long-acting nanoformulated antiretroviral therapy elicits potent antiretroviral and neuroprotective responses in HIV-1-infected humanized mice. *AIDS* 26: 2135–2144.

64. Zhou, T., et al. 2018. Creation of a nanoformulated cabotegravir prodrug with improved antiretroviral profiles. *Biomaterials* 151: 53–65.

65. Arainga, M., H. Su, L.Y. Poluektova, S. Gorantla, and H.E. Gendelman. 2016. HIV-1 cellular and tissue replication patterns in infected humanized mice. *Science and Reports* 6: 23513.

66. Arainga, M., et al. 2017. A mature macrophage is a principal HIV-1 cellular reservoir in humanized mice after treatment with long acting antiretroviral therapy. *Retrovirology* 14: 17.

67. Gnanadhas, D.P., et al. 2017. Autophagy facilitates macrophage depots of sustained-release nanoformulated antiretroviral drugs. *The Journal of Clinical Investigation* 127: 857–873.

68. Dixon, S.J., et al. 2012. Ferroptosis: An iron-dependent form of nonapoptotic cell death. *Cell* 149: 1060–1072.

69. Stockwell, B.R., et al. 2017. Ferroptosis: A regulated cell death nexus linking metabolism, redox biology, and disease. *Cell* 171: 273–285.

70. Zou, Y., et al. 2019. A GPX4-dependent cancer cell state underlies the clear-cell morphology and confers sensitivity to ferroptosis. *Nature Communications* 10: 1617.

71. Marshall, K.R., et al. 2005. The human apoptosis-inducing protein AMID is an oxidoreductase with a modified flavin cofactor and DNA binding activity. *Journal of Biological Chemistry* 280: 30735–30740.

72. Zhang, Y., et al. 2019. Imidazole ketone erastin induces ferroptosis and slows tumor growth in a mouse lymphoma model. *Cell Chemical Biology* 26: 623-633.e9.

73. Dixon, S.J., and B.R. Stockwell. 2019. The hallmarks of ferroptosis. *Annual Review of Cancer Biology* 3: 35–54.

74. Doll, S., et al. 2019. FSP1 is a glutathione-independent ferroptosis suppressor. *Nature*. https://doi.org/10.1038/s41586-019-1707-0.

75. Genovese, G., et al. 2014. Clonal hematopoiesis and blood-cancer risk inferred from blood DNA sequence. *New England Journal of Medicine* 371: 2477–2487.

76. Jaiswal, S., et al. 2014. Age-related clonal hematopoiesis associated with adverse outcomes. *New England Journal of Medicine* 371: 2488–2498.

77. Yokoyama, A., et al. 2019. Age-related remodelling of oesophageal epithelia by mutated cancer drivers. *Nature* 565: 312–317.

78. Martincorena, I., et al. 2018. Somatic mutant clones colonize the human esophagus with age. *Science* 362: 911–917.

79. Martincorena, I. et al. 2015. Tumor evolution. High burden and pervasive positive selection of somatic mutations in normal human skin. *Science* 348, 880–886.

80. Martincorena, I., and P.J. Campbell. 2015. Somatic mutation in cancer and normal cells. *Science* 349: 1483–1489.

81. Blokzijl, F., et al. 2016. Tissue-specific mutation accumulation in human adult stem cells during life. *Nature* 538: 260–264.

82. Maxwell, J.R., et al. 2015. Differential roles for interleukin-23 and interleukin-17 in intestinal immunoregulation. *Immunity* 43: 739–750.

83. Murthy, A.K., C.N. Dubose, J.A. Banas, J.J. Coalson, and B.P. Arulanandam. 2006. Contribution of polymeric immunoglobulin receptor to regulation of intestinal inflammation in dextran sulfate sodium-induced colitis. *Journal of Gastroenterology and Hepatology* 21: 1372–1380.

84. Lee, J.S., et al. 2015. Interleukin-23-independent IL-17 production regulates intestinal epithelial permeability. *Immunity* 43: 727–738.

85. Nagahama, Y., et al. 2018. Regnase-1 controls colon epithelial regeneration via regulation of mTOR and purine metabolism. *Proceedings of the National academy of Sciences of the United States of America* 115: 11036–11041.

86. Alexandrov, L.B., et al. 2013. Signatures of mutational processes in human cancer. *Nature* 500: 415–421.

87. Kumar, P., et al. 2016. Intestinal interleukin-17 receptor signaling mediates reciprocal control of the gut microbiota and autoimmune inflammation. *Immunity* 44: 659–671.

88. Cao, A.T., S. Yao, B. Gong, C.O. Elson, and Y. Cong. 2012. Th17 cells upregulate polymeric Ig receptor and intestinal IgA and contribute to intestinal homeostasis. *The Journal of Immunology* 189: 4666–4673.

89. Howell, K.J., et al. 2018. DNA methylation and transcription patterns in intestinal epithelial cells from pediatric patients with inflammatory bowel diseases differentiate disease subtypes and associate with outcome. *Gastroenterology* 154: 585–598.

90. McGovern, D.P., S. Kugathasan, and J.H. Cho. 2015. Genetics of inflammatory bowel diseases. *Gastroenterology* 149: 1163–1176.

91. Fujii, M., et al. 2018. Human intestinal organoids maintain self-renewal capacity and cellular diversity in niche-inspired culture condition. *Cell Stem Cell* 23: 787–793.

92. Fujii, M., et al. 2016. A colorectal tumor organoid library demonstrates progressive loss of niche factor requirements during tumorigenesis. *Cell Stem Cell* 18: 827–838.

93. Planell, N., et al. 2013. Transcriptional analysis of the intestinal mucosa of patients with ulcerative colitis in remission reveals lasting epithelial cell alterations. *Gut* 62: 967–976.

94. Vanhove, W., et al. 2015. Strong upregulation of AIM2 and IFI16 inflammasomes in the mucosa of patients with active inflammatory bowel disease. *Inflammatory Bowel Diseases* 21: 2673–2682.

95. Baillie, J.K. 2014. Targeting the host immune response to fight infection. *Science* 344: 807–808.

96. Warfield, K.L., et al. 2019. Lack of selective resistance of influenza A virus in presence of host-targeted antiviral, UV-4B. *Science and Reports*. https://doi.org/10.1038/s41598-019-43030-y.

97. Vercauteren, K., et al. 2016. Targeting a host-cell entry factor barricades antiviral-resistant HCV variants from on-therapy breakthrough in human-liver mice. *Gut*. https://doi.org/10.1136/gutjnl-2014-309045.

98. Merkulova, M., et al. 2015. Mapping the H+ (V)-ATPase interactome: Identification of proteins involved in trafficking, folding, assembly and phosphorylation. *Science and Reports* 5: 1–15.

99. Miles, A.L., S.P. Burr, G.L. Grice, and J.A. Nathan. 2017. The vacuolar-ATPase complex and assembly factors, TMEM199 and CCDC115, control HIF1α prolyl hydroxylation by regulating cellular Iron levels. *eLife* 6: 1–28.

100. Jansen, J.C., et al. 2016. CCDC115 deficiency causes a disorder of Golgi homeostasis with abnormal protein glycosylation. *American Journal of Human Genetics* 98: 310–321.

101. Jansen, J.C., et al. 2016. TMEM199 deficiency is a disorder of golgi homeostasis characterized by elevated aminotransferases, alkaline phosphatase, and cholesterol and abnormal glycosylation. *American Journal of Human Genetics* 98: 322–330.

102. Lu, S., T. Sung, N. Lin, R.T. Abraham, and B.A. Jessen. 2017. Lysosomal adaptation: how cells respond to lysosomotropic compounds. *PLoS One* 12, e0173771.

103. Ballabio, A., et al. 2011. TFEB links autophagy to lysosomal biogenesis. *Science* 332: 1429–1433.

104. Skehel, J.J., and D.C. Wiley. 2000. Receptor binding and membrane fusion in virus entry: The influenza hemagglutinin. *Annual Review of Biochemistry* 69: 531–569.

105. Carr, C.M., and P.S. Kim. 1993. A spring-loaded mechanism for the conformational change of influenza hemagglutinin. *Cell* 73: 823–832.
106. Schuberth-Wagner, C., et al. 2015. A conserved histidine in the RNA sensor RIG-I controls immune tolerance to N1–2′O-methylated Self RNA. *Immunity* 43: 41–52.
107. Lepore, M. et al. 2017. Functionally diverse human T cells recognize non-microbial antigens presented by MR1. *eLife* 6, 1–22.
108. Irvine, D.J., M.A. Purbhoo, M. Krogsgaard, and M.M. Davis. 2002. Direct observation of ligand recognition by T cells. *Nature* 419: 845–849.
109. Maciocia, P.M., et al. 2017. Targeting the T cell receptor β-chain constant region for immunotherapy of T cell malignancies. *Nature Medicine* 23: 1416–1423.
110. Castejon, O.J., A. Castellano, G.J. Arismendi, and Z. Medina. 2005. The inflammatory reaction in human traumatic oedematous cerebral cortex. *Journal of Submicroscopic Cytology and Pathology* 37: 43–52.
111. Lee, S.C., G.R. Moore, G. Golenwsky, and C.S. Raine. 1990. Multiple sclerosis: A role for astroglia in active demyelination suggested by class II MHC expression and ultrastructural study. *Journal of Neuropathology and Experimental Neurology* 49: 122–136.
112. Bao, X.R. et al. 2016. Mitochondrial dysfunction remodels one-carbon metabolism in human cells. *eLife* 5, e10575.
113. Quirós, P.M., et al. 2017. Multi-omics analysis identifies ATF4 as a key regulator of the mitochondrial stress response in mammals. *Journal of Cell Biology* 216: 2027–2045.
114. Viader, A., et al. 2013. Aberrant Schwann cell lipid metabolism linked to mitochondrial deficits leads to axon degeneration and neuropathy. *Neuron* 77: 886–898.
115. Harding, H.P., et al. 2003. An integrated stress response regulates amino acid metabolism and resistance to oxidative stress. *Molecular Cell* 11: 619–633.
116. Wek, R.C., H.Y. Jiang, and T.G. Anthony. 2006. Coping with stress: EIF2 kinases and translational control. *Biochemical Society Transactions* 34: 7–11.
117. Chen, J.J., and I.M. London. 1995. Regulation of protein synthesis by heme-regulated eIF-2α kinase. *Trends in Biochemical Sciences* 20: 105–108.
118. Kafina, M.D., and B.H. Paw. 2017. Intracellular iron and heme trafficking and metabolism in developing erythroblasts. *Metallomics* 9: 1193–1203.
119. Baker, M.J., et al. 2014. Stress-induced OMA1 activation and autocatalytic turnover regulate OPA1-dependent mitochondrial dynamics. *EMBO Journal* 33: 578–593.
120. Ehses, S., et al. 2009. Regulation of OPA1 processing and mitochondrial fusion by m-AAA protease isoenzymes and OMA1. *Journal of Cell Biology* 187: 1023–1036.
121. Adjibade, P., et al. 2017. DDX3 regulates endoplasmic reticulum stress-induced ATF4 expression. *Science and Reports* 7: 13832.
122. Chou, A., et al. 2017. Inhibition of the integrated stress response reverses cognitive deficits after traumatic brain injury. *Proceedings of the National academy of Sciences of the United States of America* 114: E6420–E6426.
123. Das, I., et al. 2015. Preventing proteostasis diseases by selective inhibition of a phosphatase regulatory subunit. *Science* 348: 239–242.
124. Liao, M., et al. 2007. Impaired dexamethasone-mediated induction of tryptophan 2,3-dioxygenase in heme-deficient rat hepatocytes: Translational control by a hepatic eIF2α kinase, the heme-regulated inhibitor. *Journal of Pharmacology and Experimental Therapeutics* 323: 979–989.
125. Ran, F.A., et al. 2015. In vivo genome editing using Staphylococcus aureus Cas9. *Nature* 520: 186–191.
126. Friedland, A.E., et al. 2015. Characterization of Staphylococcus aureus Cas9: A smaller Cas9 for all-in-one adeno-associated virus delivery and paired nickase applications. *Genome Biology* 16: 257.
127. Kleinstiver, B.P., et al. 2015. Engineered CRISPR-Cas9 nucleases with altered PAM specificities. *Nature* 523: 481–485.
128. Schlossarek, S., G. Mearini, and L. Carrier. 2011. Cardiac myosin-binding protein C in hypertrophic cardiomyopathy: Mechanisms and therapeutic opportunities. *Journal of Molecular and Cellular Cardiology* 50: 613–620.

129. Capmany, G., A. Taylor, P.R. Braude, and V.N. Bolton. 1996. The timing of pronuclear formation, DNA synthesis and cleavage in the human 1-cell embryo. *Molecular Human Reproduction* 2: 299–306.

130. Kim, D., et al. 2015. Digenome-seq: Genome-wide profiling of CRISPR–Cas9 off-target effects in human cells. *Nature Methods* 12: 237–243.

131. Kim, D., S. Kim, S. Kim, J. Park, and J.S. Kim. 2016. Genome-wide target specificities of CRISPR–Cas9 nucleases revealed by multiplex Digenome-seq. *Genome Research* 26: 406–415.

132. Robinson, J.T., et al. 2011. Integrative genomics viewer. *Nature Biotechnology* 29: 24–26.

133. Tang, L., et al. 2017. CRISPR/Cas9-mediated gene editing in human zygotes using Cas9 protein. *Molecular Genetics and Genomics* 292: 525–533.

134. Liang, P., et al. 2015. CRISPR/Cas9-mediated gene editing in human tripronuclear zygotes. *Protein & Cell* 6: 363–372.

135. Tu, Z., et al. 2017. Promoting Cas9 degradation reduces mosaic mutations in non-human primate embryos. *Science and Reports* 7: 42081.

136. Hashimoto, M., Y. Yamashita, and T. Takemoto. 2016. Electroporation of Cas9 protein/sgRNA into early pronuclear zygotes generates non-mosaic mutants in the mouse. *Developmental Biology* 418: 1–9.

137. Garanto, A. et al. 2013. Unexpected CEP290 mRNA splicing in a humanized knock-in mouse model for Leber congenital amaurosis. *PLoS One* 8, e79369.

138. LUXTURNA (voretigene neparvovec-rzyl, package insert) (Spark Therapeutics, 2017).

139. Tsai, S.Q., et al. 2015. GUIDE-seq enables genome-wide profiling of off-target cleavage by CRISPR-Cas nucleases. *Nature Biotechnology* 33: 187–197.

140. Food and Drug Administration. Policy for Diagnostics Testing in Laboratories Certified to Perform High Complexity Testing Under CLIA Prior to Emergency Use Authorization for Coronavirus Disease-2019 During the Public Health Emergency (US Food and Drug Administration, 2020).

141. Chen, J.S., et al. 2018. CRISPR–Cas12a target binding unleashes indiscriminate single-stranded DNase activity. *Science* 360: 436–439.

142. Chiu, C. 2018. Cutting-edge infectious disease diagnostics with CRISPR. *Cell Host & Microbe* 23: 702–704.

143. Gootenberg, J.S., et al. 2017. Nucleic acid detection with CRISPR-Cas13a/C2c2. *Science* 356: 438–442.

144. Myhrvold, C., et al. 2018. Field-deployable viral diagnostics using CRISPR-Cas13. *Science* 360: 444–448.

145. Li, S.-Y., et al. 2018. CRISPR–Cas12a-assisted nucleic acid detection. *Cell Discovery* 4: 20.

146. Centers for Disease Control and Prevention. Real-time RT–PCR Panel for Detection 2019-nCoV (US Centers for Disease Control and Prevention, 2020); https://www.cdc.gov/corona virus/2019-ncov/lab/rt-pcr-detection-instructions.html.

147. Bosch, I. et al. 2017. Rapid antigen tests for dengue virus serotypes and Zika virus in patient serum. *Science Translational Medicine* 9, eaan1589. https://doi.org/10.1126/scitranslmed.aan 1589.

148. Popowitch, E.B., S.S. O'Neill, and M.B. Miller. 2013. Comparison of the Biofire FilmArray RP, Genmark eSensor RVP, Luminex xTAG RVPv1, and Luminex xTAG RVP fast multiplex assays for detection of respiratory viruses. *Journal of Clinical Microbiology* 51: 1528–1533. https://doi.org/10.1128/JCM.03368-12.

149. Du, Y., et al. 2017. Coupling sensitive nucleic acid amplification with commercial pregnancy test strips. *Angewandte Chemie International Edition in English* 56: 992–996. https://doi.org/ 10.1002/anie.201609108.

150. Wang, D., et al. 2002. Microarray-based detection and genotyping of viral pathogens. *Proceedings of the National academy of Sciences of the United States of America* 99: 15687–15692. https://doi.org/10.1073/pnas.242579699.

151. Houldcroft, C.J., M.A. Beale, and J. Breuer. 2017. Clinical and biological insights from viral genome sequencing. *Nature Reviews Microbiology* 15: 183–192. https://doi.org/10.1038/nrm icro.2016.182.

152. Palacios, G., et al. 2007. Panmicrobial oligonucleotide array for diagnosis of infectious diseases. *Emerging Infectious Diseases* 13: 73–81. https://doi.org/10.3201/eid1301.060837.
153. Kulesa, A., J. Kehe, J.E. Hurtado, P. Tawde, and P.C. Blainey. 2018. Combinatorial drug discovery in nanoliter droplets. *Proceedings of the National academy of Sciences of the United States of America* 115: 6685–6690. https://doi.org/10.1073/pnas.1802233115.
154. Chertow, D.S. 2018. Next-generation diagnostics with CRISPR. *Science* 360: 381–382. https://doi.org/10.1126/science.aat4982.
155. Kocak, D.D., and C.A. Gersbach. 2018. From CRISPR scissors to virus sensors. *Nature* 557: 168–169. https://doi.org/10.1038/d41586-018-04975-8.
156. Bordi, L., et al. 2020. Differential diagnosis of illness in patients under investigation for the novel coronavirus (SARS-CoV-2), Italy, February 2020. *Eurosurveillance Weekly* 25: 2000170. https://doi.org/10.2807/1560-7917.ES.2020.25.8.2000170.
157. Hassibi, A., et al. 2018. Multiplexed identification, quantification and genotyping of infectious agents using a semiconductor biochip. *Nature Biotechnology* 36: 738–745. https://doi.org/10.1038/nbt.4179.
158. Dunbar, S.A. 2006. Applications of Luminex xMAP technology for rapid, high-throughput multiplexed nucleic acid detection. *Clinica Chimica Acta* 363: 71–82. https://doi.org/10.1016/j.cccn.2005.06.023.
159. Jinek, M., et al. 2012. A programmable dual-RNA-guided DNA endonuclease in adaptive bacterial immunity. *Science* 337: 816–821.
160. Komor, A.C., Y.B. Kim, M.S. Packer, J.A. Zuris, and D.R. Liu. 2016. Programmable editing of a target base in genomic DNA without double-stranded DNA cleavage. *Nature* 533: 420–424.
161. Gaudelli, N.M., et al. 2017. Programmable base editing of A*T to G*C in genomic DNA without DNA cleavage. *Nature* 551: 464–471.
162. Yeh, W.-H. et al. 2020. In vivo base editing restores sensory transduction and transiently improves auditory function in a mouse model of recessive deafness. *Science Translational Medicine* 12, eaay9101.
163. Yeh, W.H., H. Chiang, H.A. Rees, A.S.B. Edge, and D.R. Liu. 2018. In vivo base editing of post-mitotic sensory cells. *Nature Communications* 9: 2184.
164. McGrath, E., et al. 2019. Targeting specificity of APOBEC-based cytosine base editor in human iPSCs determined by whole genome sequencing. *Nature Communications* 10: 5353.
165. Zuo, E., et al. 2019. Cytosine base editor generates substantial off-target single-nucleotide variants in mouse embryos. *Science* 364: 289–292.
166. Jin, S., et al. 2019. Cytosine, but not adenine, base editors induce genome-wide off-target mutations in rice. *Science* 364: 292–295.

Chapter 5
Application of CRISPR-Based Technology in Plant Gene Editing and Agricultural Engineering

Ziheng Zhang, Ping Wang, and Ji-Long Liu

Introduction

Crop improvement breeding is an important way of dealing with the food security challenges brought about by rapid population growth, and it is regarded as an important scientific issue. For thousands of years, plant domestication has occurred mainly through the accumulation of favourable mutations in natural variation, which is a long and completely uncontrollable process. To obtain more excellent traits, over the past century, new mutations through chemical or radiation mutagenesis have been created, and a series of varieties with improved shapes have been produced. Identifying the mutation of interest, however, is a long and laborious process, and has seriously delayed the speed of breeding.

With the advancement of biotechnology in recent decades, the methods and approaches of plant genetic breeding have been expanded. Somatic cell fusion technology makes it possible to transfer genes between unrelated species and produces a series of crop varieties. With the development of genetic engineering, the introduction of specific genes into crops could be realised. This provides the possibility of transferring genes from other species to crops in a more precise manner. In recent years, CRISPR has developed rapidly at an unprecedented speed, and has become popular for its high specificity and high efficiency to modify specific locus, paving the way for targeted gene editing and breeding in crops.

The implantation of CRISPR/Cas9 gene editing technology comes at a huge price in the process of plant species change, including the reduction in time and

Z. Zhang
School of Life Science and Technology, ShanghaiTech University, Shanghai, China

P. Wang
University Library, ShanghaiTech University, Shanghai, China

J.-L. Liu (✉)
School of Life Science and Technology, ShanghaiTech University, Shanghai, China

© The Author(s), under exclusive license to Springer Nature Singapore Pte Ltd. 2022
Z. Zhang et al. (eds.), *CRISPR*,
https://doi.org/10.1007/978-981-16-8504-0_5

manpower and material resources. It also provides more diversified breeding strategies, including base editing. In this chapter, we introduce some of the masterpieces of the past few years, including the use of CRISPR and variants to achieve gene editing and breeding in plants.

Machine generated keywords: plant, rice, crop, wheat, maize, protoplast, substitution, seed, allele, adenine, transgenic, deaminase, promoter, induction, trait.

5.1 CRISPR-Based Gene Editing in Plants

Machine generated keywords: plant, rice, protoplast, wheat, substitution, deaminase, adenine, transgenic, adenine base, crop, abe, point mutation, editor, promoter, base editor.

Efficient DNA-free genome editing of bread wheat using CRISPR/Cas9 ribonucleoprotein complexes.

https://doi.org/10.1038/ncomms14261

Abstract-Summary

We describe an efficient genome editing method for bread wheat using CRISPR/Cas9 ribonucleoproteins (RNPs).

Deep sequencing reveals that the chance of off-target mutations in wheat cells is much lower in RNP mediated genome editing than in editing with CRISPR/Cas9 DNA.

Because no foreign DNA is used in CRISPR/Cas9 RNP mediated genome editing, the mutants obtained are completely transgene free.

This method may be widely applicable for producing genome edited crop plants and has a good prospect of being commercialized.

Introduction

CRISPR/Cas9 DNA constructs are delivered into plant cells by Agrobacterium tumefaciens mediated T-DNA transfer or biolistic bombardment, become expressed, cleave target sites and produce mutations [1].

During this process, there is a strong possibility that the CRISPR/Cas9 constructs are integrated into the plant genome [2].

In response to this concern, substantial efforts are being made to optimize CRISPR/Cas9 mediated genome editing with the aim of avoiding transgene integration and off-target mutations.

We showed that transient expression of CRISPR/Cas9 DNA or RNA (TECCDNA or TECCRNA) in wheat resulted in efficient genome editing with significantly reduced transgene integration [2].

Results

As the first step in our work, we tested if CRISPR/Cas9 RNPs may cleave targeted genomic sites and induce mutations in wheat protoplasts using the gw2-sgRNA that we previously found to be highly active on the TaGW2 gene [2].

In a parallel protoplast transfection experiment involving the expression of Cas9 and gw2-sgRNA from a plasmid DNA construct (pGE-TaGW2) [2], the mutagenesis frequencies for TaGW2-B1 and -D1 were 41.2% and 35.6%, respectively, whereas that for off-target editing of TaGW2-A1 was 30.8%.

Two days after the bombardment, we isolated genomic DNA samples from 100 pooled immature embryos that had been treated with either gw2-RNPs or pGE-TaGW2, and mutations in TaGW2-A1, -B1 and -D1 were analysed by deep amplicon sequencing, respectively.

gw2-RNPs also resulted in a much higher ratio of on-target to off-target editing events than pGE-TaGW2.

Discussion

We demonstrated that CRISPR/Cas9 RNPs delivered into immature embryo cells of bread wheat by particle bombardment were effective in performing targeted genome editing.

The different steps of our method and the technical details involved in each step may be useful for genome editing studies in other crops using CRISPR/Cas9 RNPs.

Although reason for the very low level of off-target mutations in RNP mediated genome editing remains to be determined, we speculate that avoidance of CRISPR/Cas9 transgene integration, fast cellular degradation of the RNPs, and shortened functional time of CRISPR/Cas9 may be involved.

Genome editing using CRISPR/Cas9 RNPs may provide a general method for reducing unintended off-target cleavage.

We established an efficient and specific CRISPR/Cas9 RNP mediated genome editing method for bread wheat.

Methods

Cas9 protein (1 μg) and sgRNA (1 μg) were mixed with the purified target DNA (100–150 ng) in Cas9 reaction buffer (20 mM HEPES, pH 7.5, 150 mM KCl, 10 mM $MgCl_2$, 0.5 mM DTT) in a total volume of 20 μl, followed by digestion at 37 °C for 0.5–1 h. The digested products were immediately separated on a 2% agarose gel, and cleavage activity was measured by the amount of digested products over the total amount of input target DNA.

For each shot, Cas9 protein (2 μg) and sgRNA (2 μg) were premixed in Cas9 reaction Buffer (20 mM HEPES, pH 7.5, 150 mM KCl, 10 mM $MgCl_2$, 0.5 mM DTT) in a total volume of 10 μl and incubated at 25 °C for 10 min.

Additional information

How to cite this article: Liang, Z. and others Efficient DNA-free genome editing of bread wheat using CRISPR/Cas9 ribonucleoprotein complexes.

Publisher's note: Springer Nature remains neutral with regard to jurisdictional claims in published maps and institutional affiliations.

Acknowledgements

A machine generated summary based on the work of Liang, Zhen; Chen, Kunling; Li, Tingdong; Zhang, Yi; Wang, Yanpeng; Zhao, Qian; Liu, Jinxing; Zhang, Huawei; Liu, Cuimin; Ran, Yidong; Gao, Caixia 2017 in Nature Communications.

Precise base editing in rice, wheat and maize with a Cas9-cytidine deaminase fusion.

https://doi.org/10.1038/nbt.3811

Main

We used two fusion proteins: nCas9-PBE (plant base editor) and dCas9-PBE.

Base-editing in protoplasts of the respective plants was assessed by sequencing 100,000–270,000 reads per locus.

Neither point mutations nor indels were detected among the 87 plants transformed with pH-dCas9-PBE.

An experiment still ongoing with different maize genotypes may explain why only this C to T substitution was recovered in transgenic plants when several other substitutions were found in base-edited maize protoplasts.

NCas9-PBE can accomplish efficient and site-specific C to T base editing in rice, wheat and maize.

In these plants, it has a deamination window covering 7 bases of the protospacer and produces virtually no indel mutations.

Our findings, together with base-editing reported in rice [3–5], suggest that base-editing is more efficient than TILLING and homologous-recombination-mediated generation of point mutations in crops.

Methods

The fusion cistrons APOBEC1-XTEN-n/dCas9-UGI were amplified using the primer set BamHI-F and Bsp1047I-R, and the PCR products were digested with BamHI and Bsp1047I and placed downstream of the Ubi-1 promoter in the plasmid pJIT163-Ubi-GFP by replacing the GFP coding sequence, which generated the plasmids pnCas9-PBE and pdCas9-PBE.

The construction of sgRNA expression vectors under the control of the promoters OsU3, TaU6 and ZmU3 for rice, wheat and maize, respectively, was performed as previously described [6–8].

The APOBEC1-XTEN-n/dCas9-UGI fragment was fused to StuI and SacI-digested pHUE411 (for rice) [9] or pBUE411 (for maize) [9] with the primer set Gibson-F and Gibson-R using the Gibson cloning method [10], yielding the vectors pHUE411/BUE411-APOBEC1-XTEN-n/dCas9-UGI.

The resultant constructs (pH/B-n/dCas9-PBE) were used to transform the rice cultivar Nipponbare or the maize inbred Zong31 through Agrobacterium-mediated transformation (see below).

T7E1, PCR-RE assays and Sanger sequencing were conducted as described previously [6–8] to identify the rice, wheat or maize mutants harboring C to T conversions in the target regions.

Acknowledgements

A machine generated summary based on the work of Zong, Yuan; Wang, Yanpeng; Li, Chao; Zhang, Rui; Chen, Kunling; Ran, Yidong; Qiu, Jin-Long; Wang, Daowen; Gao, Caixia 2017 in Nature Biotechnology.

A CRISPR–Cpf1 system for efficient genome editing and transcriptional repression in plants.

https://doi.org/10.1038/nplants.2017.18

Abstract-Summary

BV3L6 (As) and Lachnospiraceae bacterium ND2006 (Lb) in plants, using a dual RNA polymerase II promoter expression system.

LbCpf1 generated biallelic mutations at nearly 100% efficiency at four independent sites in rice T0 transgenic plants.

Our data suggest promising applications of CRISPR–Cpf1 for editing plant genomes and modulating the plant transcriptome.

Methods

Polyethylene glycol transformation of rice protoplasts with T-DNA vectors was carried out according to our previously published protocol [11].

Rice stable transformation was conducted as published previously [11].

Genomic DNA was extracted from transformed rice protoplasts or transgenic lines using the CTAB method [12].

To generate a full-length target sequence with 250 bp paired-end Illumina reads, the paired-ends were joined by single reads using flash software [13].

[Section 3]

Such distal cleavage allows previously mutated sequences to be severed repeatedly, promoting homology-dependent repair (HDR); (4) repetitive cleavage, coupled with extensive processing of staggered 5′ DNA ends, may also promote large chromosomal deletions; (5) the Cpf1 crRNA length (~43 nt) is less than half that of Cas9, making it more suitable for multiplexed genome editing and packaging into viral vectors.

To develop and test Cpf1 in plants, we focused on AsCpf1 and LbCpf1, both of which showed DNA cleavage activity in human cells [14].

We tested this dual Pol II promoter system for targeting six sites in three rice genes (OsPDS, OsDEP1 and OsROC5), and we assayed mutation frequencies in protoplasts.

We accessed target specificity of LbCpf1 by testing whether the enzyme is capable of tolerating mismatches between on-target DNA and the crRNA protospacer sequence.

We next tested whether we could generate rice plants with Cpf1-induced mutations.

Acknowledgements

A machine generated summary based on the work of Tang, Xu; Lowder, Levi G.; Zhang, Tao; Malzahn, Aimee A.; Zheng, Xuelian; Voytas, Daniel F.; Zhong, Zhaohui; Chen, Yiyi; Ren, Qiurong; Li, Qian; Kirkland, Elida R.; Zhang, Yong; Qi, Yiping 2017 in Nature Plants.

Targeted base editing in rice and tomato using a CRISPR-Cas9 cytidine deaminase fusion.

https://doi.org/10.1038/nbt.3833

Main

Rice-codon optimized Cas9 ($Cas9^{Os}$) [15] was mutated to produce the D10A mutant $nCas9^{Os}$ or D10A and H840A mutant $dCas9^{Os}$ and fused with PmCDA1 codon-optimized for Arabidopsis thaliana to create $dCas9^{Os}$-$PmCDA1^{At}$ and $nCas9^{Os}$-$PmCDA1^{At}$, which were transfected into rice calli by the Agrobacterium tumefaciens method.

The C287T gene mutation, which results in an A96V amino acid substitution, endows rice plants with resistance to the herbicide imazamox (IMZ).

Previously [16], spontaneous resistance mutations W548C/L were observed regardless of Target-AID treatment at a frequency of 1.56%, but $nCas9^{Os}$-$PmCDA1^{At}$ induced 3.41% IMZ tolerance.

T_0 lines edited with DELLA-targeted $nCas9^{At}$-$PmCDA1^{Hs}$ or $nCas9^{At}$-$PmCDA1^{At}$ contained indels or C-to-T or C-to-G substitutions in the sgRNA-targeted sequence.

Two DELLA-targeted T_1 plants, line number 1_7 from $nCas9^{At}$-$PmCDA1^{Hs}$ and line number 1_13 from $nCas9^{At}$-$PmCDA1^{At}$, contained substitutions of two amino acids (PL to LV), but PCR analysis did not detect kanamycin resistance marker gene (NPT II) fragments in these lines.

Methods

The Arabidopsis codon–optimized coding sequence of PmCDA1 was synthesized by Eurofins Genomics (Tokyo, Japan) and inserted after d/nCas9 with the same linker peptide, as described previously [17].

CAPS analysis was performed using primers FTIP1e AID-F_HindII and FTP1e AID-R_XhoI to amplify a 1,054-bp fragment consisting of 1,034 bp genomic region and 20 bp primer sequence, which was digested with PvuII and subjected to electrophoresis.

For the tomato study, potential off-target genomic regions were selected by CCTop (http://crispr.cos.uni-heidelberg.de/index.html) [18], and the top off-target site for DELLA, as well as its on-target site, were analyzed in two independent T_0 transgenic lines.

The fragment containing the target region (0.6–2 kb) was first PCR-amplified from extracted genomic DNA using the first primer pair.

Colonies on the LB agar plates were picked, and their insert sequences were determined by the Sanger method using target-specific primers.

Acknowledgements

A machine generated summary based on the work of Shimatani, Zenpei; Kashojiya, Sachiko; Takayama, Mariko; Terada, Rie; Arazoe, Takayuki; Ishii, Hisaki; Teramura, Hiroshi; Yamamoto, Tsuyoshi; Komatsu, Hiroki; Miura, Kenji; Ezura, Hiroshi; Nishida, Keiji; Ariizumi, Tohru; Kondo, Akihiko 2017 in Nature Biotechnology.

Expanded base editing in rice and wheat using a Cas9-adenosine deaminase fusion.

https://doi.org/10.1186/s13059-018-1443-z

Abstract-Summary

Nucleotide base editors in plants have been limited to conversion of cytosine to thymine.

We describe a new plant adenine base editor based on an evolved tRNA adenosine deaminase fused to the nickase CRISPR/Cas9, enabling A•T to G•C conversion at frequencies up to 7.5% in protoplasts and 59.1% in regenerated rice and wheat plants.

Extended:

These findings, together with previously described plant substitution systems [19–22], extend the application of base editing to the majority of codons and now provides feasible opportunities for significant in vivo mutagenesis studies and trait improvement in plants.

Background

Base editing is a unique genome editing system that creates precise and highly predictable nucleotide substitutions at genomic targets without requiring DSBs, or donor DNA templates, or depending on NHEJ and HDR [23].

Base editing is more efficient than HDR-mediated base pair substitution, and produces fewer undesirable mutations in the target locus [24].

The most commonly used base editing systems, such as BE3 [25], BE4 [26], Targeted-AID [27], and dCpf1-BE [28], use Cas9 or Cpf1 variants to recruit cytidine deaminases that exploit DNA mismatch repair pathways and generate specific C to T substitutions.

Results

These results demonstrate that the plant ABE system can induce A to G conversions in rice, and that the presence of three NLS at the C-terminus of nCas9 maximizes editing efficiency.

The PABE-7 base editing construct, together with the esgRNA, induces A to G substitutions efficiently and with high fidelity at multiple loci in rice and wheat.

This is the first report of producing C2186R substitution of resistant rice plants using genome editing tools.

We also used the plant ABE system to generate base-edited plants in wheat by targeting TaDEP1 and TaGW2 genes.

These results support that the plant ABE system is effective in inducing specific point mutations in rice and wheat in a highly specific and precise manner without causing other genomic modifications.

Discussion

We adapted and optimized a plant ABE system (fusion of an evolved tRNA adenosine deaminase with nuclease-inactivated CRISPR/Cas9) to efficiently and specifically achieve targeted conversion of adenine to guanine in crop plants.

This is the first report of achieving wheat A to G base-edited plants and herbicide-resistant rice plants with the plant ABE system.

The herbicide-resistant rice plants harboring the C2186R substitution in OsACC was also obtained, indicating this plant ABE system is a reliable tool for achieving targeted base editing in crop plants.

The plant ABE system combined with the plant C to T base editing system by ligating sgRNA with different aptamers (MS2, PP7, COM, and boxB) [29, 30] could achieve simultaneous A to G and C to T changes, and could be used to correct point mutations related to important agronomic traits.

Conclusions

We describe here an efficient plant base-editing system that induces precise A•T to G•C substitutions across a broad range of endogenous genomic loci.

Methods

To construct vectors PABE-1 to PABE-7, the tRNA editing deaminase ecTadA, ecTadA*, 32aa linker, and nCas9 (D10A) sequences were codon-optimized for cereal plants, and synthesized commercially (GENEWIZ, Suzhou, China).

Plasmid DNA (10 μg per construct) was introduced into the desired protoplasts by PEG-mediated transfection, the mean transformation efficiency being 45–60% by flow cytometry (FCM).

At 60 h post-transfection, the protoplasts were collected to extract genomic DNA for deep amplicon sequencing and T7E1 and PCR restriction enzyme digestion assays (PCR-RE assays; see below).

Genomic DNA extracted from the desired protoplast samples at 60 h post-transfection was used as template.

The amplicon sequencing was repeated three times for each target site, using genomic DNA extracted from three independent protoplast samples.

T7E1 and PCR-RE assays and Sanger sequencing were performed to identify rice and wheat mutants with A to G conversions in target regions, as described previously [6, 7].

Acknowledgements

A machine generated summary based on the work of Li, Chao; Zong, Yuan; Wang, Yanpeng; Jin, Shuai; Zhang, Dingbo; Song, Qianna; Zhang, Rui; Gao, Caixia 2018 in Genome Biology.

Plant gene editing through de novo *induction of meristems.*
   https://doi.org/10.1038/s41587-019-0337-2

Abstract-Summary

Plant gene editing is typically performed by delivering reagents such as Cas9 and single guide RNAs to explants in culture.

We report two methods to generate gene-edited dicotyledonous plants through de novo meristem induction.

Developmental regulators and gene-editing reagents are delivered to somatic cells of whole plants.

The de novo induction of gene-edited meristems sidesteps the need for tissue culture and promises to overcome a bottleneck in plant gene editing.

Main

Meristem identity is dictated, in part, by developmental regulators (DRs); in Arabidopsis thaliana, they include WUSCHEL (WUS), SHOOT MERISTEMLESS (STM) and MONOPTEROS (MP) [31].

Because plant cells are totipotent and can be transdifferentiated into other cell types, ectopic expression of specific DR combinations in somatic cells has the potential to induce meristems.

In A. thaliana, for example, meristem-like structures are generated when WUS and STM or the irrepressible variant of MP (ΔMP) are expressed in leaf cells [32, 33].

Expression of specific DRs in plant somatic cells can induce other developmental programs.

In monocots such as maize and sorghum, expression of maize (Zea mays) Wuschel2 (Wus2) and Baby Boom (Bbm) promotes somatic cells to form embryos that develop into whole plants [34–36].

We report that the concomitant expression of DRs and gene-editing reagents creates transgenic and gene-edited shoots through de novo meristem induction.

Results

Having demonstrated that Fast-TrACC can be used to induce meristems, we next tested different combinations of DRs expressed from promoters of varying strengths to determine the best combination for the production of full plants.

To creating transgenic plants, we wanted to determine whether Fast-TrACC could generate gene-edited meristems and plants that transmit targeted mutations to their progeny.

Based on these collective data, we conclude that co-delivery of DRs and gene-editing reagents can produce shoots with mutations and that these mutations can be transmitted to the next generation.

Based on our ability to generate luciferase-positive shoots, we concluded that ectopic delivery of DRs can create transgenic meristems and shoots on soil-grown plants.

We conclude that shoots with targeted gene edits can be generated on soil-grown plants through the use of DRs in combination with gene-editing reagents.

Discussion

Tissue culture is also crucial for the success of plant gene-editing applications.

Although regeneration of edited or transformed plant cells by tissue culture has been successful in some species and genotypes, it can be time-consuming and often introduces unintended changes to the genome and epigenome of regenerated plants [37, 38].

We report two approaches in which DRs and gene-editing reagents can be combined effectively to create transgenic and gene-edited plants.

In the second strategy, gene-edited shoots were induced on soil-grown plants, eliminating the need for aseptic culture.

Both approaches are remarkably efficient, requiring no more than 5 to 15 plants to create multiple gene-edited shoots.

To the de novo induction of transgenic and gene-edited meristems shown here, others have had some success in creating transgenic plants by delivering transgenes directly to existing meristems in, for example, the monocot wheat [39].

Online Methods

Before incubating with seedlings, the culture is again centrifuged and resuspended to OD_{600} within the range of 0.10 to 0.18 in a 50:50 (v/v) mix of AB:MES salts and half-strength Murashige and Skoog (MS) liquid plant growth medium (half-strength MS salt supplemented with 0.5% sucrose (w/v), pH 5.5).

Plants were immediately inoculated with A. tumefaciens cultures at the wound sites using syringes and 31 G needles.

Plants were observed for shoot formation at cut sites 38–48 d after inoculation.

Potato plants (Solanum tuberosum, Ranger Russet) were propagated aseptically on full strength MS medium (3% sucrose, 0.75% plant agar, pH 5.6–5.7) for 2 weeks before inoculation with A. tumefaciens.

Plants were immediately inoculated with A. tumefaciens cultures at the wound sites using syringes with 31 G needles as described above.

Samples with a single unique sequence modification in > 90% of reads were considered homozygous for the observed mutation.

Acknowledgements

A machine generated summary based on the work of Maher, Michael F.; Nasti, Ryan A.; Vollbrecht, Macy; Starker, Colby G.; Clark, Matthew D.; Voytas, Daniel F. 2019 in Nature Biotechnology.

Prime genome editing in rice and wheat.

https://doi.org/10.1038/s41587-020-0455-x

Abstract-Summary

We adapted prime editors for use in plants through codon, promoter, and editing-condition optimization.

Regenerated prime-edited rice plants were obtained at frequencies of up to 21.8%.

Main

The RT domain uses a nicked genomic DNA strand as a primer for the synthesis of an edited DNA flap templated by an extension on the pegRNA.

When we examined prime editing of endogenous genes by the PPE-CaMV system, we found that PPE-CaMV generated the desired 6-bp deletion with 5.8% efficiency at the OsCDC48-T1 site, and the desired G-to-A substitution with 0.3% efficiency at the OsCDC48-T2 site.

Editing efficiency decreases with increasing length of insertion or deletion, but PPE can install small DNA insertions and deletions into genomic sites with useful efficiencies.

The PPEs reported here can efficiently produce a wide variety of edits at genomic sites in rice and wheat, especially when pegRNA designs and editing conditions are optimized.

Although plant prime editing, like mammalian prime editing, is less efficient than base editors for making transition point mutations [40], our study shows that PPEs can generate transversions, mixtures of different substitutions, insertions, and deletions.

Methods

To construct the pegRNA expression vectors, sgRNAs were amplified using primer sets containing the target sgRNA sequences in the forward primer and the PBS + RT sequences in the reverse primer, and cloned into the OsU3-sgRNA or TaU6-sgRNA vectors [6, 7].

To construct the binary vector pH-nCas9-PPE3 for Agrobacterium-mediated rice transformation, PPE, pegRNA and nicking sgRNA expression cassettes were cloned into the pHUE411 backbone [9] by using a ClonExpressII One Step Cloning Kit (Vazyme).

The targeted sequences were amplified with specific primers, and the amplicons were purified with an EasyPure PCR Purification Kit (TransGen Biotech) and quantified with a NanoDrop 2000 spectrophotometer (Thermo Fisher Scientific).

In the first round PCR, the target region was amplified from protoplast genomic DNA with site-specific primers.

Amplicon sequencing was repeated three times for each target site using genomic DNA extracted from three independent protoplast samples.

Acknowledgements

A machine generated summary based on the work of Lin, Qiupeng; Zong, Yuan; Xue, Chenxiao; Wang, Shengxing; Jin, Shuai; Zhu, Zixu; Wang, Yanpeng; Anzalone, Andrew V.; Raguram, Aditya; Doman, Jordan L.; Liu, David R.; Gao, Caixia 2020 in Nature Biotechnology.

Single-nucleotide editing for zebra3 and wsl5 phenotypes in rice using CRISPR/Cas9-mediated adenine base editors.

https://doi.org/10.1007/s42994-020-00018-x

Abstract-Summary

The CRISPR/Cas9-mediated base editing technology can efficiently generate point mutations in the genome without introducing a double-strand break (DSB) or supplying a DNA donor template for homology-directed repair (HDR).

Adenine base editors (ABEs) were used for rapid generation of precise point mutations in two distinct genes, OsWSL5, and OsZEBRA3 (Z3), in both rice protoplasts and regenerated plants.

The precisely engineered point mutations were stably inherited to subsequent generations.

The ABE vectors and the method from this study could be used to simultaneously generate point mutations in multiple target genes in a single transformation and serve as a useful base editing tool for crop improvement as well as basic studies in plant biology.

Extended:

We achieved successful and precise A → G base editing at two gene loci in rice using two versions of adenine base editor, ABE7.10 and ABE7.9.

Introduction

To install precise point mutations and to delete or introduce desired DNA sequences, we highly rely on HDR-mediated precise genome editing.

CRISPR/Cas-mediated base editing systems have recently been developed to precisely generate point mutations in a genome (Komor and others 25; Nishida and others 41; Gaudelli and others 42).

The mutant plant possesses a single base substitution (T → C) in the third exon, causing a missense mutation (serine to proline at amino acid 542), with mutant z3 plants exhibiting a phenotype of alternating transverse dark-green/light-green stripes in the leaves and growth stunting.

We report the development of a plant base editing system based on E. coli TadA-derived adenine base editors, demonstrate its utility for single nucleotide mutations and their germline transmission, and provide evidence for the associated mutant phenotypes.

This system allows one to generate non-HDR-based indel-free and multiplexed base change mutations and to readily generate transgene-free, single base edited plants through selfing or backcrossing.

Materials and Methods

The PTG containing sgRNA for both WSL5 and Z3 genes was inserted similarly into the BsaI digested pK-ABE7.10 and pK-ABE7.9 vector.

Mannitol was replaced with enzyme solution (1.5% Cellulase R10, 0.75% Macerozyme R10, 0.5 M mannitol, 10 mM MES at pH 5.7, 1 mM $CaCl_2$, 5 mM β-marcaptoethanol, and 0.1% BSA) and the rice tissues were digested for 5–8 h in dark with gentle shaking.

After incubation for 1 h at room temperature, W5 solution was removed by centrifugation and protoplasts were re-suspended in MMG solution (4 mM MES, 0.6 M Mannitol, 15 mM $MgCl_2$) to a final concentration of 5×10^6 cells ml^{-1}.

The protoplasts and DNA were gently mixed, and 1 ml of freshly prepared PEG solution (0.6 M Mannitol, 100 mM $CaCl_2$, and 40% PEG4000) was added slowly.

Results

From these ABE-positive transgenic plants, the target loci (WSL5 and Z3) were amplified by PCR.

While we initially intended to obtain individual plants with both sites mutated, due to the low efficiency of base editing at the WSL5 locus, we were unable to obtain stable transgenic lines with both types of mutations.

Out of 84 mutant T_0 plants, not a single plant exhibited unintended proximal base editing.

We then assumed if the active base editor could induce mutation in the other targeted locus in the single mutant plants in T_1 generation, i.e., checking WSL5 locus in Z3 mutant plant and vice versa.

We further looked at whether the successful targeted base editing alters the phenotype of rice plants.

Among all T_1 plants derived from both wsl5 and z3 monoallelic T_0 mutant lines, we obtained 21% of plants devoid of the base-editor.

Discussion

We achieved successful and precise A → G base editing at two gene loci in rice using two versions of adenine base editor, ABE7.10 and ABE7.9.

Although earlier studies showed ABEs applicability in rice base-editing, they have not reported the evidence of germline transmission of the mutation (Hua and others 43; Li and others 44; Yan and others 45).

One of the reasons to select WSL5 and Z3 as the target genes in our study to demonstrate ABE effectivity was that the base editing could be translated into a detectable phenotype.

Studies on plant ABE did not report the generation of base editor free mutants (Hua and others 43, 46; Li and others 44; Yan and others 45).

Our study demonstrates that ABEs can be used to generate precise and heritable base change mutations in plants and to rapidly develop edited, yet transgene-free, crops.

Acknowledgements

A machine generated summary based on the work of Molla, Kutubuddin A.; Shih, Justin; Yang, Yinong 2020 in aBIOTECH.

5.2 Application of CRISPR-Based Technology in Crop Breeding

Machine generated keywords: plant, crop, rice, seed, allele, trait, variety, elite, maize, yield, induction, line, breeding, seven, field.

De novo *domestication of wild tomato using genome editing.*

https://doi.org/10.1038/nbt.4272

Main

T_1 seeds were harvested from plant 3, which showed an oval fruit phenotype, indicative of successful editing of the ovate locus, and determinate growth habit, indicative of loss of function of the self-pruning gene.

The four edited genes were SELF-PRUNING (SP), OVATE (O), FRUIT WEIGHT 2.2 (FW2.2) and LYCOPENE BETA CYCLASE (CycB).

We carried out a second round of genome engineering to generate multiplex-edited plants in which the selection of target genes focused on modulating fruit size (FW2.2, FAS) and number (MULT), as well as nutritional value (LYCOPENE BETA CYCLASE, CycB).

Introduction of a loss-of-function MULT allele into cultivated tomato resulted in a higher number of fruits per truss, conferring enhanced yield [41].

We demonstrate that simultaneous CRISPR–Cas9 editing of six genes resulted in modification of fruit number, size, shape, nutrient content and plant architecture in a single generation and within a single transformation experiment.

Methods

Seeds were germinated on semi-solid MS medium (supplemented with 0.6 g/L agar) and incubated at $25 \pm 1\ °C$ in the dark for 4 d. After this period, seeds were transferred and maintained at 25 ± 1 °C under long-day conditions (16 h light/ 8 h dark) with 45 μmol photons $m^{-2} s^{-1}$ PAR irradiance.

After acclimation, these plants were grown in a greenhouse at 30 °C/26 °C day/night temperature and 60–75% ambient relative humidity, 11.5 h/13 h (winter/summer) photoperiod, sunlight 250–350 μmol photons $m^{-2} s^{-1}$ PAR irradiance, attained by a reflecting mesh (Aluminet, Polysack Indústrias Ltda, Leme, Piracicaba, SP, Brazil), and automatic irrigation four times a day for fruit set.

Seeds from transformed plants were germinated in 350-mL pots with a 1:1 mixture of commercial potting mix Basaplant (Base Agro, Artur Nogueira, SP, Brazil) and expanded vermiculite supplemented with $1 g L^{-1}$ 10:10:10 NPK and $4 g L^{-1}$ dolomite limestone ($MgCO_3$ plus $CaCO_3$).

Acknowledgements

A machine generated summary based on the work of Zsögön, Agustin; Čermák, Tomáš; Naves, Emmanuel Rezende; Notini, Marcela Morato; Edel, Kai H; Weinl, Stefan; Freschi, Luciano; Voytas, Daniel F; Kudla, Jörg; Peres, Lázaro Eustáquio Pereira 2018 in Nature Biotechnology.

Clonal seeds from hybrid rice by simultaneous genome engineering of meiosis and fertilization genes.

https://doi.org/10.1038/s41587-018-0003-0

Abstract-Summary

Clonal propagation through seeds would enable self-propagation of F_1 hybrids.

We fixed the heterozygosity of F_1 hybrid rice by multiplex CRISPR–Cas9 genome editing of the REC8, PAIR1 and OSD1 meiotic genes to produce clonal diploid gametes and tetraploid seeds.

We demonstrated that editing the MATRILINEAL (MTL) gene (involved in fertilization) could induce formation of haploid seeds in hybrid rice.

We combined fixation of heterozygosity and haploid induction by simultaneous editing of all four genes (REC8, PAIR1, OSD1 and MTL) in hybrid rice and obtained plants that could propagate clonally through seeds.

Main

The self-fertilization of MiMe plants doubles the ploidy at each generation.

The results demonstrate that recombinational haploid plants can be generated by self-fertilization of hybrid varieties.

Taking these results together, the diploid progeny of Fix plants displayed the same ploidy, the same heterozygous genotype, and a phenotype similar to that of the parent Fix plants, implying that Fix is able to produce clonal seeds and fix the heterozygosity of F_1 hybrid rice.

Our findings revealed that hybrids can be self-pollinated to produce true-breeding progeny through seeds by targeted editing of four endogenous genes in a rice F_1 hybrid variety.

When the four genes were simultaneously mutated in hybrid rice by genome editing, the Fix plants displayed a similar reduced fertility.

Although the fertility was reduced mainly because of the MTL mutation, the Fix plant was able to produce clonal seeds with the same ploidy and heterozygous genotype.

Methods

Genomic DNA was extracted from approximately 100 mg rice leaf tissue via the cetyltrimethylammonium bromide (CTAB) method.

Approximately 2 cm^2 leaf tissue was chopped using a new razor blade for 2 to 3 min in 1 mL LB01 buffer (15 mM Tris, 2 mM disodium EDTA, 0.5 mM spermine tetrahydrochloride, 80 mM KCl, 20 mM NaCl, 0.1% (v/v) Triton X-100, 15 mM β-mercaptoethanol, pH 7.5, filtered through a 0.22-μm filter).

The 150-bp paired-end reads were generated by Illumina Hiseq2500 at an average depth of approximately 30-fold coverage for each sample.

The raw paired-end reads were first filtered to create clean data using NGSQC-toolkit v2.3.3 [42].

Clean reads of each accession were aligned against the rice reference genome (IRGSP 1.0) using the software SOAPaligner (soap version 2.21) [43] with parameters "-m 200 -x 1000 -l 35 -s 42 -v 5" and "--p 8".

Acknowledgements

A machine generated summary based on the work of Wang, Chun; Liu, Qing; Shen, Yi; Hua, Yufeng; Wang, Junjie; Lin, Jianrong; Wu, Mingguo; Sun, Tingting; Cheng, Zhukuan; Mercier, Raphael; Wang, Kejian 2019 in Nature Biotechnology.

One-step genome editing of elite crop germplasm during haploid induction.

https://doi.org/10.1038/s41587-019-0038-x

Abstract-Summary

We co-opted the aberrant reproductive process of haploid induction (HI) [44–47] to induce edits in nascent seeds of diverse monocot and dicot species.

Our method, named HI-Edit, enables direct genomic modification of commercial crop varieties.

HI-Edit was tested in field and sweet corn using a native haploid-inducer line [45] and extended to dicots using an engineered CENH3 HI system [48].

Edited plants could be used in trait testing and directly integrated into commercial variety development.

Main

To investigate whether matl pollen can edit commercial inbred germplasm, the non-haploid inducer inbred maize line NP2222 was stably transformed with a binary vector expressing Cas9 and a guide RNA (gRNA) targeting the site of the reference matl-1 mutation in the MATL gene [49].

HI-Edit therefore provides direct evidence in support of the hypothesis that maize haploids derive from post-fertilization genome elimination, analogous to the mechanism of CENH3 [48].

To validate HI-Edit for breeding targets, the maize inbred NP2222 was stably transformed with constructs expressing Cas9 and gRNAs targeting VRS1-LIKE HOMEOBOX PROTEIN duplicate genes VLHP1 (GRMZM2G104204) and VLHP2 (GRMZM2G062244) [50], or GRAIN WEIGHT2 duplicate genes ZmGW2-1 (GRMZM2G170088) and ZmGW2-2 (GRMZM2G007288) [51].

We used HI-Edit to enable genome editing of monocots (maize and wheat) and dicots (Arabidopsis) by both maternal and paternal HI mechanisms.

Optimization of crossing methods and Cas9 expression by gamete-preferential promoters (for example DUO1 [52] for maternal HI, and EGG CELL1 (EC1) [53] for paternal HI) would avoid unwanted editing during vegetative development and may boost efficiency by delivering Cas9 ribonucleoprotein at fertilization.

Methods

The other three edited plants failed to amplify Taqman assays detecting Cas9 and the selectable marker PHOSPHOMANNOSE ISOMERASE (PMI) which was present in the transformation construct; all three were haploids by flow cytometry.

The T0 transformants were assayed for modification of the GW2 target sequences by qPCR, and efficiently edited single-copy events were selected for crossing with the haploid-inducer line RWKS [49].

Tissues from seedling haploid plants were sampled and assayed to determine if editing had occurred at the target sites (ZmVLHP1, ZmVLHP2, ZmGW2-1 and ZmGW2-2), if the transgene (Cas9, PMI) is present and the status of the HI gene MATL (wild type) and matl (haploid inducing allele from the male).

The markers we used to detect the edited haploids were as follows: a '0' score for ZmCENH3 indicates a haploid because the maternal genome has been lost (ZmCENH3 is homozygous in the inducer plant).

Acknowledgements

A machine generated summary based on the work of Kelliher, Timothy; Starr, Dakota; Su, Xiujuan; Tang, Guozhu; Chen, Zhongying; Carter, Jared; Wittich, Peter E.; Dong, Shujie; Green, Julie; Burch, Erin; McCuiston, Jamie; Gu, Weining; Sun, Yuejin; Strebe, Tim; Roberts, James; Bate, Nic J.; Que, Qiudeng 2019 in Nature Biotechnology.

Broad-spectrum resistance to bacterial blight in rice using genome editing.

https://doi.org/10.1038/s41587-019-0267-z

Abstract-Summary

Oryzae (Xoo), secretes one or more of six known transcription-activator-like effectors (TALes) that bind specific promoter sequences and induce, at minimum, one of the three host sucrose transporter genes SWEET11, SWEET13 and SWEET14, the expression of which is required for disease susceptibility.

We used CRISPR–Cas9-mediated genome editing to introduce mutations in all three SWEET gene promoters.

Editing was further informed by sequence analyses of TALe genes in 63 Xoo strains, which revealed multiple TALe variants for SWEET13 alleles.

Paddy trials showed that genome-edited SWEET promoters endow rice lines with robust, broad-spectrum resistance.

Main

TALes from Xoo are injected by a type III secretion system into plant cells and recognize effector-binding elements (EBEs) in cognate SWEET host gene promoters, which results in induction of SWEET genes and production of sugars that enable disease susceptibility in rice [54, 55].

EBE alleles of SWEET11 that are not recognized by PthXo1 are collectively referred to as the recessive resistance gene xa13.

Most indica rice varieties carry a SWEET13 allele that contains four adenines in the EBE for PthXo2, and rice lines carrying this allele are susceptible to PthXo2-dependent strains [56].

A rare exception is the recessive resistance allele xa25, which contains three adenines in the EBE for SWEET13 in the indica cultivar Minghui 63, conferring resistance to strains that depend solely on PthXo2 [57].

Building on the understanding of TALes and their SWEET gene target EBEs, we developed a strategy to use CRISPR–Cas9 genome editing to engineer broad and durable resistance to bacterial blight in rice.

[Section 2]

An assessment of the diversity of the major TALes in extant strains of Xoo was obtained by analyzing Xoo genome sequence data and virulence assays on rice lines with SWEET promoter polymorphisms.

In African lineages, the two known major TALes of Xoo are TalC and TalF, which each induce SWEET14 using non-overlapping EBEs [58, 59].

Diversity of TALes From Infection Trials

To estimate the prevalence and variation of TALes in geographically diverse Xoo strains, a collection of 105 Xoo strains was screened against the japonica rice variety Kitaake and derivative Kitaake lines carrying mutations in the three SWEET genes normally targeted by TALes.

Ten of 105 Xoo strains were not virulent on the japonica reference line Kitaake.

These strains may be solely dependent on PthXo2 for induction of the IR24 indica allele of SWEET13, as the EBEs in the other SWEET gene promoters are the same in Kitaake and IR24.

[Section 4]

Both PthXo2B and PthXo2C were predicted to bind to the same EBE (TATAAAG-CACCACAACTCCCTTC) within the SWEET13 promoter of Kitaake and other japonica varieties.

Xoo strains with PthXo2 alone are incompatible on Kitaake-derived lines owing to the inability to induce the japonica allele of SWEET13 [56].

The presence of pthXo2B or pthXo2C could explain the virulence phenotypes of the seven deviant strains on Kitaake and the double-edited Kitaake line 52-1.

To test this possibility, the candidate gene for PthXo2B from PXO61 was cloned and introduced into strain PXO99^{A}ME2 (hereafter ME2), which is not pathogenic on any of the tested rice lines owing to a null mutation in the sole major TALe gene pthXo1 of PXO99^A (ref [60]).

[Section 5]

We tested whether stacking multiple mutations in the EBEs of three SWEET promoters known to be targeted by major TALes would provide resistance to most, if not all, Xoo strains in a single rice line.

[Section 6]

If the combined SWEET editing events tested here do not affect the normal physiological function of the genes, we predict that the same combinations would not impair agronomic traits in rice mega varieties, and would thus be useful for rice breeding programs.

A range of lines with different insertions, deletions and substitutions in SWEET11 (n = 6), SWEET13 (n = 8) and SWEET14 (n = 7) were selected, assuming that larger EBE differences will be less likely to match TALe variation in pathogen populations.

Distinct combinations of these 21 SWEET variants were distributed across nine rice lines.

Discussion

Most Xoo strains, if not all, target some combination of six short EBE sequences present in the promoters of the three SWEET genes.

We applied multiplexed CRISPR–Cas9 genome editing to systematically interfere with SWEET gene induction at all known major TALe EBEs and engineer Kitaake rice resistant to all currently known strains of Xoo.

These seven Asian strains retained virulence on the first-generation of genome-edited EBE lines in the Kitaake background, indicating that the variant strains either carry another major TALe targeting an undefined susceptibility gene or a variant major TALe targeting SWEET11, SWEET13 or SWEE14 at an alternate EBE.

By using a combination of systematic analyses of diverse Xoo strains, an understanding of SWEET genes and genome editing, we were able to engineer broad-spectrum resistance in Kitaake and two mega varieties IR64 and Ciherang-Sub1.

Methods

Agrobacterium strains containing the respective CRISPR constructs were used for genome editing in Kitaake, IR64 and Ciherang-Sub1.

Thirty plants for each of the 19 IR64 T2 lines were phenotyped for resistance to Xoo strains PXO339, PXO99 and PXO86, and target promoter regions of the candidate lines were sequenced to confirm mutations.

Seeds of T2 plants were bulked (from 30 plants per line), phenotyped for Xoo resistance and analyzed for agronomic traits as described below.

Seeds from 13 individual T3 plants were advanced to T4 on the basis of three criteria: (1) mutation type, (2) consistent resistance to the three Xoo strains and (3) seed count.

Eighteen Cas9-free T1 plants were further analyzed for resistance and amplicon sequencing as described above up to the T3 generation.

Acknowledgements

A machine generated summary based on the work of Oliva, Ricardo; Ji, Chonghui; Atienza-Grande, Genelou; Huguet-Tapia, José C.; Perez-Quintero, Alvaro; Li, Ting; Eom, Joon-Seob; Li, Chenhao; Nguyen, Hanna; Liu, Bo; Auguy, Florence; Sciallano, Coline; Luu, Van T.; Dossa, Gerbert S.; Cunnac, Sébastien; Schmidt, Sarah M.; Slamet-Loedin, Inez H.; Vera Cruz, Casiana; Szurek, Boris; Frommer, Wolf B.; White, Frank F.; Yang, Bing 2019 in Nature Biotechnology.

Rapid customization of Solanaceae fruit crops for urban agriculture.

https://doi.org/10.1038/s41587-019-0361-2

Abstract-Summary

Cultivation of crops in urban environments might reduce the environmental impact of food production [61–64].

Lack of available land in cities and a need for rapid crop cycling, to yield quickly and continuously, mean that so far only lettuce and related 'leafy green' vegetables are cultivated in urban farms [65].

New fruit varieties with architectures and yields suitable for urban farming have proven difficult to breed [61, 65].

Application of our strategy using one-step CRISPR–Cas9 genome editing restructured vine-like tomato plants into compact, early yielding plants suitable for urban agriculture.

Main

There is much interest in urban agriculture of fruits and berries, but developing crop varieties suitable to the restrictive growth parameters of urban agriculture farming systems requires considerable modification.

While these sp sp5g 'double-determinate' genotypes are rapid cycling and productive when grown at high density in greenhouses and fields [66], even smaller plants that produce commercially viable yields would be better suited to urban agriculture.

These results demonstrate that high-performing, triple-determinate, small-fruited tomato varieties can be developed to accommodate the plant size and space restrictions of urban agriculture.

Our strategy enables rapid engineering of two Solanaceae fruit crops to the most challenging agronomic parameters of urban agriculture: rapid cycling and compact plant size.

Our CRISPR–Cas9-based approach will enable rapid modification of many other small-fruited tomato varieties into a triple-determinate growth habit by generating loss-of-function alleles of SP, SP5G and SIER in elite breeding lines.

Methods

Plants were grown in a greenhouse under long-day conditions (16 h light, 26–28 °C/8 h dark, 18–20 °C; 40–60% relative humidity) supplemented with artificial light from high-pressure sodium bulbs (~250 μmol m^{-2} s^{-1}), in the agricultural fields at Cold Spring Harbor Laboratory, Cornell Long Island Horticultural Experiment Station, Riverhead, New York, USA, and at the Gulf Coast Research and Education Center, Wimauma, Florida, USA.

Quantification data on tomato and groundcherry shoots and inflorescences were obtained from the individual plants grown in greenhouses and fields at Cold Spring Harbor Laboratory.

The yield trial for Sweet100 sp, sp sp5g and sp sp5g sler was conducted on plants grown in the fields of Cornell Long Island Horticultural Experiment Station, Riverhead, New York, USA (9 August 2019).

Plants were grown under long-day conditions (16 h light, 26–28 °C/8 h dark, 18–20 °C; ambient humidity) with artificial light from an LED (475 μmol m^{-2} s^{-1}) with 4,000 K color temperature at Cornell University, Ithaca, New York.

Acknowledgements

A machine generated summary based on the work of Kwon, Choon-Tak; Heo, Jung; Lemmon, Zachary H.; Capua, Yossi; Hutton, Samuel F.; Van Eck, Joyce; Park, Soon Ju; Lippman, Zachary B 2019 in Nature Biotechnology.

Marker-free carotenoid-enriched rice generated through targeted gene insertion using CRISPR-Cas9.

https://doi.org/10.1038/s41467-020-14981-y

Abstract-Summary

Targeted insertion of transgenes at pre-determined plant genomic safe harbors provides a desirable alternative to insertions at random sites achieved through conventional methods.

We report the use of an optimized CRISPR-Cas9-based method to achieve the targeted insertion of a 5.2 kb carotenoid biosynthesis cassette at two genomic safe harbors in rice.

These results demonstrate targeted gene insertion of marker-free DNA in rice using CRISPR-Cas9 genome editing, and offer a promising strategy for genetic improvement of rice and other crops.

Introduction

Targeted gene insertion at double-strand breaks (DSBs) in the GSHs provides a desirable alternative to conventional plant transformation methods [67].

Recent advances in genome editing technologies have enabled the induction of DSB at defined targets in a relatively simple manner, paving the way for targeted gene insertion in plants [68, 69].

Most reported examples of targeted gene insertion by CRISPR-Cas in plants are dependent on chemical selection of the inserted cassette [70–76].

The few cases of targeted insertion of marker-free DNA fragments in plants have been achieved with relatively small DNA fragments (ranging from 281 bp to 1.8 kb) [77–80].

We demonstrate the targeted insertion of a 5.2 kb marker-free DNA fragment at two GSHs in rice using CRISPR-Cas, and obtain homozygous carotenoid-enriched rice.

Results

We performed genetic segregation analysis of the T1 generation to obtain rice plants homozygous for the carotenoid cassette at Target B that lack the Cas9-gRNA module.

We did not detect any DNA sequence of the pAcc-B donor plasmid in the genome of 48A-7 besides Target B. Together, these results suggest that plant 48A-7 carries a single copy of the full-length carotenoid cassette at Target B. To assess the occurrence of off-target mutations caused by Cas9 in the process, we further analyzed the whole-genome sequencing result for 48A-7.

This indicates that the β-carotene in the seeds from 48A-7 results from the targeted insertion of the carotenoid cassette at Target B. To test whether the method of targeted insertion described above can be applied to other chromosomal locations, and to assess the frequency of insertion of the donor DNA, we performed an additional round of co-bombardment experiment at a different target site, Target C. In this experiment, we cultivated each callus separately to prevent clonal propagation.

Discussion

Conventional plant genetic engineering methods rely on the insertion of genes encoding desirable agronomic traits at random positions in the genome.

Events like this may potentially be reduced with prior knowledge of GSHs within a given genome, and the availability of a reliable tool to insert desired genes at these sites.

To identify GSHs that express transgenes at desired levels, a population of independent transgenic events, each carrying a reporter gene (such as GFP) at a distinct insertion site, can be generated and screened.

The homozygous transgenic lines expressing the reporter gene without exhibiting unfavorable agronomic traits carry the reproter gene insert at GSHs.

With targeted gene insertion as reported in this study, genes encoding multiple traits genes can be stacked at a single genetic locus, which would simplify the downstream breeding efforts.

Methods

For particle bombardment, Kitaake seeds were sterilized and germinated on the MSD medium (MS with 3% sucrose, 2 mg L^{-1} 2,4-D and 1.2% Agar, pH 5.7) under 28 °C in the dark for 7 days for initial calli induction.

Cells from each tube were resuspended in 2 mL of the WI Solution (0.5 M Mannitol, 20 mM KCl, 4 mM MES pH 5.7, 25 μg mL^{-1} carbenicillin) and kept in the dark at 25 °C for 70 h. Genomic DNA was then extracted from the protoplast cells using the CTAB-chloroform-based method.

For the T7E1 assay, genomic DNA fragments spanning various targets were amplified with the Phusion High-Fidelity DNA Polymerase System (Thermo Fisher, Waltham, MA) using primers Target-A/B/C/D/E-PCR- F and R. The PCR products were heated to 95 °C and ramped down to 25 °C over 14 min evenly to allow heteroduplex formation.

Acknowledgements

A machine generated summary based on the work of Dong, Oliver Xiaoou; Yu, Shu; Jain, Rashmi; Zhang, Nan; Duong, Phat Q.; Butler, Corinne; Li, Yan; Lipzen, Anna; Martin, Joel A.; Barry, Kerrie W.; Schmutz, Jeremy; Tian, Li; Ronald, Pamela C. 2020 in Nature Communications.

Superior field performance of waxy corn engineered using CRISPR–Cas9.

https://doi.org/10.1038/s41587-020-0444-0

Abstract-Summary

We created waxy corn hybrids by CRISPR–Cas9 editing of a waxy allele in 12 elite inbred maize lines, a process that was more than a year faster than conventional trait introgression using backcrossing and marker-assisted selection.

Main

Of these, the wx-C allele is the most widely used wx donor in modern commercial waxy hybrids.

Commercial waxy hybrids are developed by introgressing a wx mutant allele into elite inbred lines by backcrossing.

We report on the application of CRISPR–Cas9 and morphogenic genes to produce waxy deletion alleles directly in 12 elite maize inbreds and multi-location yield testing, all in three years, which is substantially faster than introgression methods.

Maize kernels homozygous for CRISPR wx-d1 alleles had the same waxy phenotype as kernels of conventional waxy lines that carry the wx-C allele introduced by trait introgression (TI-wx).

Five stiff stalk and three nonstiff stalk inbreds produced homozygous wx-d1 seeds before the planting date of the 2016 winter nursery, allowing generation of seven hybrids.

CRISPR-wx hybrids yielded approximately 5.5 bu ac^{-1} (P < 0.05) more than introgressed waxy hybrids.

Methods

To facilitate delivery of the genome editing reagents (Cas9, guide RNAs) into maize cells and regeneration of plants, morphogenic transcription factors Bbm (also called ovule development protein2 or ODP2) and Wus2 were expressed under control of the maize tissue specific promoter of the phospholipid transferase gene (Zm-PLTPpro) and auxin-responsive gene1 (Zm-Axig1pro), respectively, and the plasmids were constructed as described previously [35, 71].

DNA was extracted from leaf punches of selected BC0 and BC1 wx deletion plants as described previously [81].

In the single-location analysis, the main effect of entry type (TI-wx, CRISPR-wx) was considered as a fixed effect, and the hybrid and the interaction between entry type and hybrid were treated as random effects.

In the multi-location analysis, the main effect of entry type (TI-wx, CRISPR-wx) was considered as a fixed effect, and the hybrid and the interaction between entry type and hybrid were treated as random effects.

Acknowledgements

A machine generated summary based on the work of Gao, Huirong; Gadlage, Mark J.; Lafitte, H. Renee; Lenderts, Brian; Yang, Meizhu; Schroder, Megan; Farrell, Jeffry; Snopek, Kay; Peterson, Dave; Feigenbutz, Lanie; Jones, Spencer; St Clair, Grace; Rahe, Melissa; Sanyour-Doyel, Nathalie; Peng, Chenna; Wang, Lijuan; Young, Joshua K.; Beatty, Mary; Dahlke, Brian; Hazebroek, Jan; Greene, Thomas W.; Cigan, A. Mark; Chilcoat, N. Doane; Meeley, R. Bob 2020 in Nature Biotechnology.

References

1. Bortesi, L., and R. Fischer. 2015. The CRISPR/Cas9 system for plant genome editing and beyond. *Biotechnology Advances* 33: 41–52.
2. Zhang, Y., et al. 2016. Efficient and transgene-free genome editing in wheat through transient expression of CRISPR/Cas9 DNA or RNA. *Nature Communications* 7: 12617.
3. Lu, Y., and J.K. Zhu. 2016. *Molecular Plant*. https://doi.org/10.1016/j.molp.2016.11.013.

4. Li, J., Y. Sun, J. Du, Y. Zhao, and L. Xia. 2016. Generation of targeted point mutations in rice by a modified CRISPR/Cas9 system. *Molecular Plant*. https://doi.org/10.1016/j.molp.2016. 12.001.
5. Ren, B. et al. 2016. Sci. China Life Sci. http://engine.scichina.com/doi/10.1007/s11427-016-0406-x.
6. Shan, Q., et al. 2013. Targeted genome modification of crop plants using a CRISPR-Cas system. *Nature Biotechnology* 31: 686–688.
7. Wang, Y., et al. 2014. Simultaneous editing of three homoeoalleles in hexaploid bread wheat confers heritable resistance to powdery mildew. *Nature Biotechnology* 32: 947–951.
8. Liang, Z., K. Zhang, K. Chen, and C.J. Gao. 2014. *Genetics Genomics* 41: 63–68.
9. Xing, H.L., et al. 2014. *BMC Plant Biology* 14: 327.
10. Kurowska, M., et al. 2011. *Journal of Applied Genetics* 52: 371–390.
11. Tang, X., et al. 2016. *Molecular Plant* 9: 1088–1091.
12. Murray, M.G., and W.F. Thompson. 1980. *Nucleic Acids Research* 8: 4321–4325.
13. Magoc, T., and S.L. Salzberg. 2011. *Bioinformatics* 27: 2957–2963.
14. Zetsche, B., et al. 2015. Cpf1 is a single RNA-guided endonuclease of a class 2 CRISPR-Cas system. *Cell* 163: 759–771.
15. Mikami, M., S. Toki, and M. Endo. 2015. *Plant Molecular Biology* 88: 561–572.
16. Yu, Q., and S.B. Powles. 2014. *Pest Management Science* 70: 1340–1350.
17. Nishida, K., et al. 2016. *Science* 102: 553–563.
18. Stemmer, M., T. Thumberger, M. Del Sol Keyer, J. Wittbrodt, and J.L. Mateo. 2015. CCTop: An intuitive, flexible and reliable CRISPR/Cas9 target prediction tool. PLoS One 10: e0124633–e11.
19. Hua, K., X. Tao, F. Yuan, D. Wang, and J.K. Zhu. 2018. Precise A.T to G.C base editing in the rice genome. Molecular Plant 11:627–30.
20. Shimatani, Z., S. Kashojiya, M. Takayama, R. Terada, T. Arazoe, H. Ishii, H. Teramura, T. Yamamoto, H. Komatsu, K. Miura, et al. 2017. Targeted base editing in rice and tomato using a CRISPR-Cas9 cytidine deaminase fusion. *Nature Biotechnology* 35: 441–443.
21. Yan, F., Y. Kuang, B. Ren, J. Wang, D. Zhang, H. Lin, B. Yang, X. Zhou, and H. Zhou. 2018. Highly efficient A.T to G.C base editing by Cas9n-guided tRNA adenosine deaminase in rice. Molecular Plant 11:631–634.
22. Zong, Y., et al. 2017. Precise base editing in rice, wheat and maize with a Cas9–cytidine deaminase fusion. *Nature Biotechnology* 35: 438–440.
23. Hess, G.T., J. Tycko, D. Yao, and M.C. Bassik. 2017. Methods and applications of CRISPR-mediated base editing in eukaryotic genomes. *Molecular Cell* 68: 26–43.
24. Yang, B., X. Li, L. Lei, and J. Chen. 2017. APOBEC: From mutator to editor. *Journal of Genetics and Genomics* 44: 423–437.
25. Komor, A.C., Y.B. Kim, M.S. Packer, J.A. Zuris, and D.R. Liu. 2016. Programmable editing of a target base in genomic DNA without double-stranded DNA cleavage. *Nature* 533: 420–424.
26. Komor, A. C. et al. 2017. Improved base excision repair inhibition and bacteriophage Mu Gam protein yields C:G-to-T:A base editors with higher efficiency and product purity. Science Advances 3: eaao4774.
27. Nishida, K., T. Arazoe, N. Yachie, S. Banno, M. Kakimoto, M. Tabata, M. Mochizuki, A. Miyabe, M. Araki, K.Y. Hara, et al. 2016. Targeted nucleotide editing using hybrid prokaryotic and vertebrate adaptive immune systems. *Science* 353: 1248.
28. Li, X., Y. Wang, Y. Liu, B. Yang, X. Wang, J. Wei, Z. Lu, Y. Zhang, J. Wu, X. Huang, et al. 2018. Base editing with a Cpf1–cytidine deaminase fusion. *Nature Biotechnology* 36: 324–327.
29. Zalatan, J.G., et al. 2015. Engineering complex synthetic transcriptional programs with CRISPR RNA scaffolds. *Cell* 160: 339–350.
30. Ma, H., et al. 2016. Multiplexed labeling of genomic loci with dCas9 and engineered sgRNAs using CRISPRainbow. *Nature Biotechnology* 34: 528–530.
31. Barton, M.K. 2010. Twenty years on: The inner workings of the shoot apical meristem, a developmental dynamo. *Developmental Biology* 341: 95–113.

32. Gallois, J.-L., C. Woodward, G.V. Reddy, and R. Sablowski. 2002. Combined SHOOT MERIS-TEMLESS and WUSCHEL trigger ectopic organogenesis in Arabidopsis. *Development* 129: 3207–3217.

33. Ckurshumova, W., T. Smirnova, D. Marcos, Y. Zayed, and T. Berleth. 2014. Irrepressible MONOPTEROS/ARF5 promotes de novo shoot formation. *New Phytologist* 204: 556–566.

34. Lowe, K., et al. 2016. Morphogenic regulators Baby boom and Wuschel improve monocot transformation. *The Plant Cell* 28: 1998–2015.

35. Lowe, K. et al. 2018. Rapid genotype "independent" Zea mays L. (maize) transformation via direct somatic embryogenesis. *Vitro Cellulluar Development Biology, Plant* 54: 240–252.

36. Nelson-Vasilchik, K., J. Hague, M. Mookkan, Z.J. Zhang, and A. Kausch. 2018. Transformation of recalcitrant Sorghum varieties facilitated by Baby Boom and Wuschel2. Curr. Protoc. *Plant Biology* 3: e20076.

37. Phillips, R.L., S.M. Kaeppler, and P. Olhoft. 1994. Genetic instability of plant tissue cultures: Breakdown of normal controls. *Proceedings of the National academy of Sciences of the United States of America* 91: 5222–5226.

38. Zhang, D. et al. 2014. Tissue culture-induced heritable genomic variation in rice, and their phenotypic implications. *PloS ONE* 9: e96879.

39. Hamada, H., et al. 2017. An in planta biolistic method for stable wheat transformation. *Science and Reports* 7: 11443.

40. Anzalone, A.V., et al. 2019. Search-and-replace genome editing without double-strand breaks or donor DNA. *Nature* 576: 149–157.

41. Lippman, Z.B. et al. 2008. The making of a compound inflorescence in tomato and related nightshades. *PLoS Biology.* 6: e288.

42. Patel, R.K., and Jain, M. 2012. NGS QC Toolkit: A toolkit for quality control of next generation sequencing data. *PLoS ONE* 7, e30619.

43. Li, R., et al. 2009. SOAP2: An improved ultrafast tool for short read alignment. *Bioinformatics* 25: 1966–1967.

44. Laurie, D.A., and M.D. Bennett. 1988. The production of haploid wheat plants from wheat × maize crosses. *Theoretical and Applied Genetics* 76: 393–397.

45. Coe, E.H. 1959. A line of maize with high haploid frequency. *The American Naturalist* 93: 381–382.

46. Kasha, K.J., and K.N. Kao. 1970. High frequency haploid production in barley (Hordeum vulgare L.). *Nature* 225: 874–876.

47. Burke, L.G. et al. 1979. Maternal haploids of Nicotiana tabacum L. from seed. *Science* 206: 585.

48. Ravi, M., and S.W.L. Chan. 2010. Haploid plants produced by centromere-mediated genome elimination. *Nature* 464: 615–618.

49. Kelliher, T., et al. 2017. Matrilineal, a sperm-specific phospholipase, triggers maize haploid induction. *Nature* 542: 105–109.

50. Whipple, C.J., et al. 2011. GRASSY TILLERS1 promotes apical dominance in maize and responds to shade signals in the grasses. *Proceedings of the National academy of Sciences of the United States of America* 108: E506–E512.

51. Li, Q., et al. 2010. Relationship, evolutionary fate and function of two maize co-orthologs of rice GW2-associated with kernel size and weight. *BMC Plant Biology* 10: 143.

52. Borg, M., et al. 2011. The R2R3 MYB transcription factor DUO1 activates a male germline-specific regulon essential for sperm cell differentiation in Arabidopsis. *The Plant Cell* 23: 534–549.

53. Sprunck, S., et al. 2012. Egg cell-secreted EC1 triggers sperm cell activation during double fertilization. *Science* 338: 1093–1097.

54. White, F.F., N. Potnis, J.B. Jones, and R. Koebnik. 2009. The type III effectors of Xanthomonas. *Molecular Plant Pathology* 10: 749–766.

55. Bezrutczyk, M., et al. 2018. Sugar flux and signaling in plant-microbe interactions. *The Plant Journal* 93: 675–685.

56. Zhou, J., et al. 2015. Gene targeting by the TAL effector PthXo2 reveals cryptic resistance gene for bacterial blight of rice. *The Plant Journal* 82: 632–643.

57. Liu, Q., et al. 2011. A paralog of the MtN3/saliva family recessively confers race-specific resistance to Xanthomonas oryzae in rice. *Plant, Cell and Environment* 34: 1958–1969.

58. Streubel, J. et al. 2013. Five phylogenetically close rice SWEET genes confer TAL effector-mediated susceptibility to Xanthomonas oryzae pv. oryzae. *New Phytology* 200: 808–819.

59. Yu, Y. et al. 2011. Colonization of rice leaf blades by an African strain of Xanthomonas oryzae pv. oryzae depends on a new TAL effector that induces the rice nodulin-3 Os11N3 gene. Molecular Plant Microbe Interact 24: 1102–1113.

60. Yang, B., A. Sugio, and F.F. White. 2006. Os8N3 is a host disease-susceptibility gene for bacterial blight of rice. *Proceedings of the National academy of Sciences of the United States of America* 103: 10503–10508.

61. Benke, K., and B. Tomkins. 2017. Future food-production systems: Vertical farming and controlled-environment agriculture. *Sustainability Science Practice Policy* 13: 13–26.

62. Pearson, L.J., L. Pearson, and C.J. Pearson. 2010. Sustainable urban agriculture: Stocktake and opportunities. *International Journal of Agricultural Sustainability* 8: 7–19.

63. Martellozzo, F. et al. 2014. Urban agriculture: a global analysis of the space constraint to meet urban vegetable demand. *Environment Research Letter* 9: 064025.

64. Banerjee, C., and L. Adenaeuer. 2014. Up, up and away! The economics of vertical farming. *Journal of Agriculture Studies* 2: 40–60.

65. Touliatos, D., I.C. Dodd, and M. McAinsh. 2016. Vertical farming increases lettuce yield per unit area compared to conventional horizontal hydroponics. *Food Energy Security* 5: 184–191.

66. Soyk, S., et al. 2017. Variation in the flowering gene SELF PRUNING 5G promotes day-neutrality and early yield in tomato. *Nature Genetics* 49: 162–168.

67. Yamamoto, Y., and S.A. Gerbi. 2018. Making ends meet: Targeted integration of DNA fragments by genome editing. *Chromosoma* 127: 405–420.

68. Sun, Y., J. Li, and L. Xia. 2016. Precise genome modification via sequence-specific nucleases-mediated gene targeting for crop improvement. *Frontiers in Plant Science* 7: 1928.

69. Schindele, A., A. Dorn, and H. Puchta. 2019. CRISPR/Cas brings plant biology and breeding into the fast lane. *Current Opinion in Biotechnology* 61: 7–14.

70. Li, Z., et al. 2015. Cas9-guide RNA directed genome editing in soybean. *Plant Physiology* 169: 960–970.

71. Svitashev, S., et al. 2015. *Plant Physiology* 169: 931–945.

72. Svitashev, S., C. Schwartz, B. Lenderts, J.K. Young, and A. Mark Cigan. 2016. Genome editing in maize directed by CRISPR-Cas9 ribonucleoprotein complexes. *Nature Communications* 7: 13274.

73. Begemann, M.B., et al. 2017. Precise insertion and guided editing of higher plant genomes using Cpf1 CRISPR nucleases. *Science and Reports* 7: 11606.

74. Wang, M.G., et al. 2017. Gene targeting by homology-directed repair in rice using a geminivirus-based CRISPR/Cas9 system. *Molecular Plant* 10: 1007–1010.

75. Čermák, T., N.J. Baltes, R. Čegan, Y. Zhang, and D.F. Voytas. 2015. *Genome Biology* 16: 232.

76. Lee, K., et al. 2019. CRISPR/Cas9-mediated targeted T-DNA integration in rice. *Plant Molecular Biology* 99: 317–328.

77. Shi, J., et al. 2017. ARGOS8 variants generated by CRISPR-Cas9 improve maize grain yield under field drought stress conditions. *Plant Biotechnology Journal* 15: 207–216.

78. Dahan-Meir, T., et al. 2018. Efficient in planta gene targeting in tomato using geminiviral replicons and the CRISPR/Cas9 system. *The Plant Journal* 95: 5–16.

79. Miki, D., W. Zhang, W. Zeng, Z. Feng, and J.K. Zhu. 2018. CRISPR/Cas9-mediated gene targeting in Arabidopsis using sequential transformation. *Nature Communications* 9: 1967.

80. Li, J., et al. 2016. Gene replacements and insertions in rice by intron targeting using CRISPR-Cas9. *Nature Plants* 2: 16139.

81. Zastrow-Hayes, G., et al. 2015. *Plant Genome* 8: 1–15.

Chapter 6
CRISPR Guides

Ziheng Zhang, Ping Wang, and Ji-Long Liu

Machine generated keywords: NHEJ, TALEN, knockin, ZFN, barcode, protocol, HDR, CRISPRi, sgRNA, genetically, repair template, repair, screening, screen, repression.

Considering that AI systems cannot effectively summarize the types of "protocols" paper, especially "materials" and "processes", we consider removing all summaries of "materials" and "processes" in this chapter.

https://doi.org/10.1038/nprot.2013.132

Abstract-Summary

The Cas9-sgRNA complex binds to DNA elements complementary to the sgRNA and causes a steric block that halts transcript elongation by RNA polymerase, resulting in the repression of the target gene.

We provide a protocol for the design, construction and expression of customized sgRNAs for transcriptional repression of any gene of interest.

CRISPRi provides a simplified approach for rapid gene repression within 1–2 weeks.

The method can also be adapted for high-throughput interrogation of genome-wide gene functions and genetic interactions, thus providing a complementary approach to RNA interference, which can be used in a wider variety of organisms.

Extended:

Z. Zhang
School of Life Science and Technology, ShanghaiTech University, Shanghai, China

P. Wang
University Library, ShanghaiTech University, Shanghai, China

J.-L. Liu (✉)
School of Life Science and Technology, ShanghaiTech University, Shanghai, China

We provide a protocol for the design, construction and utilization of sgRNAs for sequence-specific silencing of genes at the transcriptional level.

Introduction

We have shown in Escherichia coli [1] that dCas9, when coexpressed with an sgRNA designed with a 20-bp complementary region to any gene of interest, can efficiently silence a target gene with up to 99.9% repression.

Repression efficiency can be tuned by introducing single or multiple mismatches into the sgRNA base-pairing region or by targeting different loci along the target gene.

CRISPRi presents an efficient and specific genome-targeting platform for transcription control without altering the target DNA sequence, and it can potentially be adapted as a versatile genome regulation method in diverse organisms.

Previously, the binding specificity of the dCas9-sgRNA complex to the target DNA is determined by sgRNA-DNA base pairing and an NGG PAM motif [2].

We and others have shown that the 12-nt sequence that targets the region adjacent to the PAM site constitutes a seed region that is important for effective gene regulation [2, 3].

Anticipated results

The dCas9 and designed sgRNAs can robustly silence target genes with a 99–99.5% repression level in E. coli.

In human cells, we observed moderate but reproducible repression of gene expression using properly designed sgRNAs.

Acknowledgements

A machine generated summary based on the work of Larson, Matthew H; Gilbert, Luke A; Wang, Xiaowo; Lim, Wendell A; Weissman, Jonathan S; Qi, Lei S. 2013 in Nature Protocols.

Genome engineering using the CRISPR-Cas9 system.

https://doi.org/10.1038/nprot.2013.143

Abstract-Summary

Targeted nucleases are powerful tools for mediating genome alteration with high precision.

The RNA-guided Cas9 nuclease from the microbial clustered regularly interspaced short palindromic repeats (CRISPR) adaptive immune system can be used to facilitate efficient genome engineering in eukaryotic cells by simply specifying a 20-nt targeting sequence within its guide RNA.

We describe a set of tools for Cas9-mediated genome editing via nonhomologous end joining (NHEJ) or homology-directed repair (HDR) in mammalian cells, as well as generation of modified cell lines for downstream functional studies.

Introduction

As with other designer nuclease technologies such as ZFNs and TALENs, Cas9 can facilitate targeted DNA DSBs at specific loci of interest in the mammalian genome and stimulate genome editing via NHEJ or HDR.

Cas9 can be targeted to specific genomic loci via a 20-nt guide sequence on the sgRNA.

The only requirement for the selection of Cas9 target sites is the presence of a PAM sequence directly 3′ of the 20-bp target sequence.

The CRISPR Design Tool provides the sequences for all oligos and primers necessary for (i) preparing the sgRNA constructs, (ii) assaying target modification efficiency and (iii) assessing cleavage at potential off-target sites.

For the detection of NHEJ mutations, it is important to design primers situated at least 50 bp from the Cas9 target site to allow for the detection of longer indels.

Anticipated results

That we have not been able to detect HDR in HUES9 cells by using Cas9n with a sgRNA, which may be due to low efficiency or a potential difference in repair activities in HUES9 cells.

Acknowledgements

A machine generated summary based on the work of Ran, F Ann; Hsu, Patrick D; Wright, Jason; Agarwala, Vineeta; Scott, David A; Zhang, Feng 2013 in Nature Protocols.

Generating genetically modified mice using CRISPR/Cas-mediated genome engineering.

https://doi.org/10.1038/nprot.2014.134

Abstract-Summary

Crispr/Cas technology is a quick and efficient method of modifying the genomes of a range of organisms.

Introduction

In conventional gene-targeting methods [5], mutant mice are generated by introducing mutations through homologous recombination (HR) in mouse ES cells.

The generation of mutant mice by HR in ES cells is costly and time-consuming because gene-targeted ES cell clones need to be selected and injected into blastocysts to generate chimeric mice, which then have to be bred to generate single-gene mutant offspring, a procedure that usually takes 9–12 months.

It has been shown that Cas9 homologs from other bacteria use different PAM sequences [6, 7] and also induce targeted DNA cleavage in mammalian cells.

The desired mutations, including gene disruption, point mutation, small tag insertion or conditional allele generation, can be verified by PCR amplification of the sequence around the target site, by cloning the product into the original TA Cloning vector and by subsequently validating the mutation by sequencing.

Anticipated results

On the basis of the targeting experiments that we have done, almost all injections with only Cas9 mRNA and sgRNAs resulted in mutant alleles containing indels with high efficiency (>80% for one allele).

It may be helpful to determine the targeting efficiency at the blastocyst stage by genotyping PCR using a genome amplification kit (Sigma-Aldrich, cat no.

Acknowledgements

A machine generated summary based on the work of Yang, Hui; Wang, Haoyi; Jaenisch, Rudolf 2014 in Nature Protocols.

Genome editing in rice and wheat using the CRISPR/Cas system.

https://doi.org/10.1038/nprot.2014.157

Abstract-Summary

The Gao laboratory provides its protocol for targeted editing of crop genomes using the CRISPR/Cas system.

Introduction

The CRISPR/Cas system creates DNA DSBs at target loci to stimulate genome editing via NHEJ or HDR, just like other engineered nuclease technologies such as ZFNs and TALENs.

In the CRISPR/Cas system, the only requirement for the target site is the 20-bp target sequence preceding a 5'-NGG PAM.

The targeting specificity of the CRISPR/Cas system is only dependent on sgRNAs, which are encoded by short sequences of ~100 bp, it is possible to achieve simultaneous multiplex gene editing of plant loci by co-transforming multiple sgRNAs.

A 5'-NGG PAM sequence is required downstream of target sites for CRISPR/Cas-induced cleavage, which may limit the range of available targets.

If sgRNA activity will be detected by the PCR/restriction enzyme (PCR/RE) assay (PROCEDURE Step 29A), the restriction enzyme sites within the target sequences at the Cas9 endonuclease cutting site (3-bp upstream 5'-NGG) will facilitate detection.

Anticipated results

This protocol generates sgRNA expression vectors that can be used to target any gene of interest in rice and wheat.

Acknowledgements

A machine generated summary based on the work of Shan, Qiwei; Wang, Yanpeng; Li, Jun; Gao, Caixia 2014 in Nature Protocols.

MMEJ-assisted gene knock-in using TALENs and CRISPR-Cas9 with the PITCh systems.

https://doi.org/10.1038/nprot.2015.140

Abstract-Summary

This Protocol describes the CRIS-PITCh and TAL-PITCh systems for MMEJ-based gene targeting using CRISPR-Cas or TALENs.

Introduction

To generate corresponding microhomologies at the DNA ends of the exogenous DNA donor, the microhomologies comprising the spacer sequence of the TALEN target site on the TAL-PITCh vector is switched compared with the genomic target; i.e., the 5′ junction microhomology is on the right and the 3′ junction microhomology is on the left.

On the basis of direct comparisons at a single genomic locus, the knock-in efficiency of TAL-PITCh is ~2.5-fold higher than the efficiency of HR-assisted gene knock-in in HeLa cells [8].

For TAL-PITCh, a TALEN vector should be constructed only for the genomic target site.

As 50–60 bp of sequence can be added by synthetic oligo primers, genomic PCR is not necessary for PITCh vector construction, unlike the construction of the targeting vector for HR-mediated method.

As a negative control of PITCh knock-in, the TAL-PITCh or CRIS-PITCh vector should be transfected without co-transfecting TALEN or CRISPR-Cas9 vectors.

Anticipated results

The amount of fluorescent gene expression is determined by the expression level of the corresponding endogenous promoter.

If the targeted gene is expressed at low levels, the background frequency of random integrants will increase.

Acknowledgements

A machine generated summary based on the work of Sakuma, Tetsushi; Nakade, Shota; Sakane, Yuto; Suzuki, Ken-Ichi T; Yamamoto, Takashi 2015 in Nature Protocols.

Efficient, footprint-free human iPSC genome editing by consolidation of Cas9/CRISPR and piggyBac technologies.

https://doi.org/10.1038/nprot.2016.152

Introduction

Applications of iPSCs have been greatly expanded by the advent of genome editing, in which the genomic sequence at a target site is altered by insertion or deletion ('indel') mutations, or by introduction of precisely programmed ('knockin') modifications [9].

Including the excision-only piggyBac mutant in the transfection mix with gRNA and donor DNA permits efficient, single-step genome editing and transgene excision.

A concern of Cas9-based genome-editing strategies has been off-target mutagenesis.

Whole-genome sequencing of six individual clones isolated using this protocol from separate genome-editing experiments at three loci (TAZ, DNAJC19 and JUP) showed that Cas9 does not induce frequent off-target mutagenesis.

Whole-genome sequencing of several mutant iPSC lines generated through application of this protocol did not reveal a substantial burden of off-target mutation, although each of the six cell lines that underwent whole-genome sequencing did acquire 10–15 off-target mutations each, of which 1–3 were attributable to gRNA-related Cas9 activity.

Anticipated results

Genome editing using iPSC-Cas9-PB cells is highly efficient.

Acknowledgements

A machine generated summary based on the work of Wang, Gang; Yang, Luhan; Grishin, Dennis; Rios, Xavier; Ye, Lillian Y; Hu, Yong; Li, Kai; Zhang, Donghui; Church, George M; Pu, William T. 2016 in Nature Protocols.

Genome-scale CRISPR-Cas9 knockout and transcriptional activation screening.

https://doi.org/10.1038/nprot.2017.016

Introduction

As a proof of principle to demonstrate the CRISPR-Cas9 system's utility for screening, we constructed genome-scale CRISPR-Cas9 knockout (GeCKO) and SAM libraries to identify genes that, upon knockout or activation, confer resistance to the BRAF inhibitor vemurafenib in a melanoma cell line [10, 11].

Although specificity and efficiency will probably vary across experimental settings, false-positive sgRNAs in screens can still be mitigated by including redundant sgRNAs in the library and requiring multiple distinct sgRNAs targeting the same gene to display the same phenotype when identifying screening hits.

Screening analysis methods such as RNAi gene enrichment ranking (RIGER), redundant siRNA activity (RSA), model-based analysis of genome-wide CRISPR/Cas9 knockout (MAGeCK) and STARS typically select candidate genes with multiple enriched or depleted sgRNAs to reduce the possibility that the observed change in sgRNA distribution was due to off-target activity of a single sgRNA [12–15].

Anticipated results

SgRNAs targeting genes involved in vemurafenib resistance are enriched because they provide a proliferation advantage upon vemurafenib treatment.

Acknowledgements

A machine generated summary based on the work of Joung, Julia; Konermann, Silvana; Gootenberg, Jonathan S; Abudayyeh, Omar O; Platt, Randall J; Brigham, Mark D; Sanjana, Neville E; Zhang, Feng 2017 in Nature Protocols.

Large-scale reconstruction of cell lineages using single-cell readout of transcriptomes and CRISPR–Cas9 barcodes by scGESTALT.

https://doi.org/10.1038/s41596-018-0058-x

Abstract-Summary

We recently established a method, scGESTALT, that combines cumulative editing of a lineage barcode array by CRISPR–Cas9 with large-scale transcriptional profiling using droplet-based single-cell RNA sequencing (scRNA-seq).

The technique generates edits in the barcode array over multiple timepoints using Cas9 and pools of single-guide RNAs (sgRNAs) introduced during early and late zebrafish embryonic development, which distinguishes it from similar Cas9 lineage-tracing methods.

We provide details for (i) generating transgenic zebrafish; (ii) performing multi-timepoint barcode editing; (iii) building scRNA-seq libraries from brain tissue; and (iv) concurrently amplifying lineage barcodes from captured single cells.

scGESTALT provides a scalable platform to map lineage relationships between cell types in any system that permits genome editing during development, regeneration, or disease.

Introduction

These methods rely on the introduction of Cas9-induced stochastic mutations (insertions or deletions during DNA repair following double-strand breaks) to predefined target sites in transgenes, and use the edited sequences as lineage barcodes for clonal tracing and lineage tree reconstruction.

The scGESTALT protocol is divided into four experimental sections: (i) generation of two transgenic zebrafish lines: one for barcode expression and the other for inducible Cas9 and ubiquitous sgRNA expression; (ii) performance of multi-timepoint barcode editing using a combination of injection and induction of barcoding reagents for early and late labeling of cells, respectively; (iii) transcriptome profiling using inDrops; and (iv) concurrent extraction of lineage barcodes from captured single cells.

The data analysis consists of three major parts: (i) analysis of the transcriptome data for cell-type identification, (ii) analysis of the scGESTALT lineage barcodes for identification of editing events, and (iii) generation of a single-cell-resolved lineage tree with associated cell types.

Anticipated results

Cell relationships at the level of gene expression and lineages can then be represented on lineage trees, which can be explored to identify the timing and patterns of lineage segregation within or between tissue regions and cellular subtypes.

If the pattern is dominated by short bands at ~100–150 bp, it is advisable to avoid using the corresponding animal in scRNA-seq experiments, as it is likely that this individual's barcode edits will be predominantly large deletions, reducing the diversity of edits and the resolution of the lineage trees.

The digital gene expression matrix output (rows contain gene counts, and columns represent single cells) from the inDrops processing pipeline is used with any relevant scRNA-seq computational tool to identify cell types.

To generate scGESTALT lineage trees, the inDrops index sequences are used to match transcriptomes and lineage barcodes for the same cells.

Individual cells (identified by their inDrops index) containing each of the recovered barcodes are connected to the tips of the tree (i.e., the terminal lineage barcode sequence).

Acknowledgements

A machine generated summary based on the work of Raj, Bushra; Gagnon, James A.; Schier, Alexander F. 2018 in Nature Protocols.

Integrative analysis of pooled CRISPR genetic screens using MAGeCKFlute.

https://doi.org/10.1038/s41596-018-0113-7

Abstract-Summary

We describe how to perform computational analysis of CRISPR screens using the MAGeCKFlute pipeline.

MAGeCKFlute is distinguished from other currently available tools by its comprehensive pipeline, which contains a series of functions for analyzing CRISPR screen data.

This protocol explains how to use MAGeCKFlute to perform quality control (QC), normalization, batch effect removal, copy-number bias correction, gene hit identification and downstream functional enrichment analysis for CRISPR screens.

We also describe gene identification and data analysis in CRISPR screens involving drug treatment.

Introduction

In combination, MAGeCK and MAGeCK-VISPR allow users to perform read-count mapping, normalization and QC, as well as to identify positively and negatively selected genes in the screens.

Within MAGeCK, we demonstrate how to identify gene hits with MAGeCK RRA (Step 7A(v)) using publicly available data from a CRISPR screen performed in glioblastoma (GBM) stem-like cells (GSCs) [17].

MAGeCKFlute can be applied to remove batch effects, correct copy-number bias, identify screening hits and perform downstream functional analysis for various CRISPR screens, such as CRISPR knockout, CRISPR activation and CRISPR inhibition screens.

MAGeCKFlute can map raw reads onto a CRISPR library, normalize read counts to allow comparison between different samples, identify genes that are positively or negatively selected under the screening conditions, and explore enriched GO terms and KEGG pathways for those selected genes.

For CRISPR screens with a certain treatment, we recommend that users conduct normalization with essential genes to make the beta score comparable between treatment and control samples.

Timing

Running this protocol on the example data provided will take ~3 h on a machine with eight processing cores and at least 8 GB of RAM.

Anticipated results

The 'test' subfolder contains the main results of MAGeCK RRA or MLE, including the 'gene_summary.txt' file, which can be used in the functional analysis in Step 11.

MAGeCKFlute processes MAGeCK RRA results ('gene_summary' file) with the function FluteRRA, which compares two conditions, such as day 23 versus day 0 in Dataset 1.

Gene groups that exhibit different beta scores between treatment and control samples are colored to represent these differences.

Genes in the green group are strongly negatively selected (i.e., cells whose gene is disrupted are under-represented) in the control samples and are weakly selected (either positively or negatively) in the treatment samples.

The orange group contains genes that are weakly selected in the control samples and strongly positively selected in the treatment sample (i.e., cells whose gene is disrupted are over-represented).

Acknowledgements

A machine generated summary based on the work of Wang, Binbin; Wang, Mei; Zhang, Wubing; Xiao, Tengfei; Chen, Chen-Hao; Wu, Alexander; Wu, Feizhen; Traugh, Nicole; Wang, Xiaoqing; Li, Ziyi; Mei, Shenglin; Cui, Yingbo; Shi, Sailing; Lipp, Jesse Jonathan; Hinterndorfer, Matthias; Zuber, Johannes; Brown, Myles; Li, Wei; Liu, X. Shirley 2019 in Nature Protocols.

SHERLOCK: nucleic acid detection with CRISPR nucleases.

https://doi.org/10.1038/s41596-019-0210-2

Abstract-Summary

We have recently established a CRISPR-based diagnostic platform that combines nucleic acid pre-amplification with CRISPR–Cas enzymology for specific recognition of desired DNA or RNA sequences.

This platform, termed specific high-sensitivity enzymatic reporter unlocking (SHERLOCK), allows multiplexed, portable, and ultra-sensitive detection of RNA or DNA from clinically relevant samples.

We provide step-by-step instructions for setting up SHERLOCK assays with recombinase-mediated polymerase pre-amplification of DNA or RNA and subsequent Cas13- or Cas12-mediated detection via fluorescence and colorimetric readouts that provide results in <1 h with a setup time of less than 15 min.

Introduction

We describe a protocol for SHERLOCK nucleic acid detection (using RPA and CRISPR–Cas13) with instructions for reagent preparation, including recombinant Leptotrichia wadei Cas13 (LwaCas13a) protein expression and purification, as well as in vitro transcription (IVT) of crRNAs and sample extraction from various starting materials.

In cases in which detection of the target requires single-base distinction, such as viral and human genotyping or the detection of cancer-associated mutations in circulating nucleic acids, the specificity of Cas13 can be enhanced by the introduction of a 'synthetic mismatch' into the crRNA [19, 20].

The most basic form of SHERLOCK with LwaCas13a, an active ortholog selected from a screen of Cas13a orthologs in bacterial cells, allows detection of only a single target per reaction via a fluorescence channel or a line on a lateral flow strip.

Anticipated results

The presented protocol for expression and purification of LwaCas13a yields ~0.5–1 mg of pure protein.

Acknowledgements

A machine generated summary based on the work of Kellner, Max J.; Koob, Jeremy G.; Gootenberg, Jonathan S.; Abudayyeh, Omar O.; Zhang, Feng 2019 in Nature Protocols.

CRISPR–Cas9, CRISPRi and CRISPR-BEST-mediated genetic manipulation in streptomycetes.

https://doi.org/10.1038/s41596-020-0339-z

Abstract-Summary

New advances in genome editing techniques, particularly CRISPR-based tools, have revolutionized genetic manipulation of many organisms, including actinomycetes.

We have developed a comprehensive CRISPR toolkit that includes several variations of 'classic' CRISPR–Cas9 systems, along with CRISPRi and CRISPR-base editing systems (CRISPR-BEST) for streptomycetes.

Our CRISPR toolkit can be used to generate random-sized deletion libraries, introduce small indels, generate in-frame deletions of specific target genes, reversibly suppress gene transcription, and substitute single base pairs in streptomycete genomes.

The toolkit includes a Csy4-based multiplexing option to introduce multiple edits in a single experiment.

Introduction

The protocols facilitate highly efficient, precise, and rapid generation of desired genetic modifications in streptomycetes, showing the following advantages over the conventional double crossover–based methods, including PCR-targeting-based strategies [21]: (i) the protocol is easy to perform; in most cases only one highly efficient cloning step is required to construct a functional CRISPR plasmid; (ii) the protocol does not require genome-mapped cosmid clones of the strain to be engineered (which is typically required in the PCR-targeting approach); (iii) the protocol is relatively fast (~10 d for inactivation of a gene); (iv) the protocol is versatile and covers various genetic manipulations, including in-frame knockout, in-frame knockin, introduction of indels, single-amino-acid exchange and target gene knockdown; (v) the protocol is capable of multiplexed genome editing with a single plasmid targeting multiple genes of interest simultaneously.

Timing

The minimal timing is based on optimal conditions, that is, receiving DNA oligonucleotides from the provider the next day after ordering; receiving Sanger sequencing results from the Sanger sequencing provider the next day after sample submission; using target streptomycete strains such as S. coelicolor WT with a relatively fast growth; and having all required reagents, including competent cells, prepared beforehand.

Steps 1–7, sgRNA design: 30 minSteps 8–20, construction and validation of the desired CRISPR plasmid: 7–16 dSteps 21–33, transfer of ready-to-use CRISPR plasmids into target streptomycetes by interspecies conjugation: 2 dStep 34, evaluation of the successfully edited strains: 5–14 dSteps 35–39, (optional) plasmid curing: 7–14 dBox 2, inserting editing templates into CRISPR plasmids for in-frame deletion or foreign DNA insertions: 3–5 dBox 3, a modified electroporation-competent cell preparation protocol: 2 dBox 4, modified protocol for using the Qiagen Blood & Cell Culture Mini DNA Kit with streptomycetes: 0.5–1 d.

Anticipated results

The ssDNA bridging method for 20-nt protospacer cloning generally results in cloning efficiencies of >50% (highly dependent on the sequence of the protospacers).

The validation of the library requires whole-genome sequencing.

Within the toolkit described here, three plasmids can be used to inactivate a gene: (i) When using pCRISPR–Cas9–ScaligD, one can expect mutants bearing small indels around the DSB sites; the mutations can be validated by standard Sanger sequencing (80% editing efficiency is expected [23]). (ii) When using pCRISPR–Cas9 with editing templates, one can expect a precise in-frame deletion mutant (with a well-designed editing template, an in-frame insertion mutant can be achieved), which also can be validated by standard Sanger sequencing (95% editing efficiency is expected [23]). (iii) When using pCRISPR–cBEST, one can expect mutants with stop codon introduction; again, the mutations can be validated by standard Sanger sequencing (90% editing efficiency is expected [24]).

Acknowledgements

A machine generated summary based on the work of Tong, Yaojun; Whitford, Christopher M.; Blin, Kai; Jørgensen, Tue S.; Weber, Tilmann; Lee, Sang Yup 2020 in Nature Protocols.

BAR-Seq clonal tracking of gene-edited cells.

https://doi.org/10.1038/s41596-021-00529-x

Abstract-Summary

We detail the wet laboratory and bioinformatic BAR-Seq pipeline, a strategy for clonal tracking of cells harboring homology-directed targeted integration of a barcoding cassette.

BAR-Seq can be applied to most editing strategies, and we describe its use to investigate the clonal dynamics of human edited hematopoietic stem/progenitor cells in xenotransplanted hosts.

BAR-Seq may be applied in both basic and translational research contexts to investigate the biology of edited cells and stringently compare editing protocols at a clonal level.

Our BAR-Seq pipeline allows library preparation and validation in a few days and clonal analyses of edited cell populations in 1 week.

Extended:

BAR-Seq can be applied to bulk ex vivo cultured cells and used to track engraftment and lineage output upon transplantation.

Introduction

We developed a barcoding-based strategy (BAR-Seq) that allows clonal tracking of edited cells by means of unique molecular identifiers (barcodes, BARs) embedded in the DNA template for HDR [25].

The BAR-Seq bioinformatic pipeline also identifies the most abundant ('dominant') cell clones in the HDR-edited population, thus focusing the analyses on shared clones across samples more robustly contributing to the in vitro outgrowth or to host repopulation.

BAR-Seq enables clonal tracking of HDR-edited cells at single-cell resolution, even in the absence of reporter-expressing cassettes, by offering the possibility to investigate proliferation, differentiation, self-renewal and long-term maintenance of HDR-edited cell clones, as well as to stringently compare different editing protocols and reagents.

To verify the need for setting this threshold and estimate the appropriate value c, we strongly suggest including in the sequencing run a control amplicon derived from cells edited with the conventional non-barcoded HDR template, where no BAR should be retrieved.

Anticipated results

The BAR-Seq pipeline is designed to perform clonal tracking analyses of edited cells both in vitro and in vivo, without any limitation due to the locus, donor template or nuclease platform.

The BAR-Seq web application provides a function that estimates whether the sequenced plasmid/viral library is suitable for clonal tracking of a desired number of cells.

When applied to clonal tracking of edited HSPCs, BAR-Seq uncovered the multi-lineage and self-renewing capacity of engrafting HDR-edited HSPCs in human hematochimeric mice, with most clones shared among cell lineages long-term after transplant.

Acknowledgements

A machine generated summary based on the work of Ferrari, Samuele; Beretta, Stefano; Jacob, Aurelien; Cittaro, Davide; Albano, Luisa; Merelli, Ivan; Naldini, Luigi; Genovese, Pietro 2021 in Nature Protocols.

References

1. Qi, L.S., et al. 2013. Repurposing CRISPR as an RNA-guided platform for sequence-specific control of gene expression. *Cell* 152: 1173–1183.
2. Jinek, M., et al. 2012. A programmable dual-RNA-guided DNA endonuclease in adaptive bacterial immunity. *Science* 337: 816–821.
3. Jiang, W., D. Bikard, D. Cox, F. Zhang, and L.A. Marraffini. 2013. RNA-guided editing of bacterial genomes using CRISPR-Cas systems. *Nature Biotechnology* 31: 233–239.
4. Inoue, H., H. Nojima, and H. Okayama. 1990. High efficiency transformation of Escherichia coli with plasmids. *Gene* 96: 23–28.
5. Capecchi, M.R. 2005. Gene targeting in mice: Functional analysis of the mammalian genome for the twenty-first century. *Nature Reviews Genetics* 6: 507–512.
6. Cong, L., et al. 2013. Multiplex genome engineering using CRISPR/Cas systems. *Science* 339: 819–823.
7. Esvelt, K.M., et al. 2013. Orthogonal Cas9 proteins for RNA-guided gene regulation and editing. *Nature Methods* 10: 1116–1121.
8. Nakade, S., et al. 2014. Microhomology-mediated end-joining-dependent integration of donor DNA in cells and animals using TALENs and CRISPR/Cas9. *Nature Communications* 5: 5560.
9. Kim, H.S., J.M. Bernitz, D.F. Lee, and I.R. Lemischka. 2014. Genomic editing tools to model human diseases with isogenic pluripotent stem cells. *Stem Cells Development* 23: 2673–2686.
10. Konermann, S., et al. 2015. Genome-scale transcriptional activation by an engineered CRISPR-Cas9 complex. *Nature* 517: 583–588.
11. Shalem, O., et al. 2014. Genome-scale CRISPR-Cas9 knockout screening in human cells. *Science* 343: 84–87.
12. Doench, J.G., et al. 2016. *Nature Biotechnology* 34: 184–191.
13. Konig, R., et al. 2007. A probability-based approach for the analysis of large-scale RNAi screens. *Nature Methods* 4: 847–849.
14. Luo, B., et al. 2008. Highly parallel identification of essential genes in cancer cells. *Proceedings of the National academy of Sciences of the United States of America* 105: 20380–20385.
15. Li, W., et al. 2014. *Genome Biology* 15: 554.
16. Yin, L., et al. 2015. Multiplex conditional mutagenesis using transgenic expression of Cas9 and sgRNAs. *Genetics* 200: 431–441.
17. Toledo, C.M., et al. 2015. Genome-wide CRISPR-Cas9 screens reveal loss of redundancy between PKMYT1 and WEE1 in glioblastoma stem-like cells. *Cell Reports* 13: 2425–2439.
18. Hart, T., M. Chandrashekhar, M. Aregger, Z. Steinhart, K.R. Brown, G. MacLeod, et al. 2015. High-resolution CRISPR screens reveal fitness genes and genotype-specific cancer liabilities. *Cell* 163: 1515–1526.
19. Gootenberg, J.S., et al. 2017. Nucleic acid detection with CRISPR-Cas13a/C2c2. *Science* 356: 438–442.
20. Gootenberg, J.S., et al. 2018. Multiplexed and portable nucleic acid detection platform with Cas13, Cas12a, and Csm6. *Science* 360: 439–444.
21. Gust, B., G.L. Challis, K. Fowler, T. Kieser, and K.F. Chater. 2003. PCR-targeted Streptomyces gene replacement identifies a protein domain needed for biosynthesis of the sesquiterpene soil odor geosmin. *Proceedings of the National academy of Sciences of the United States of America* 100: 1541–1546.
22. MacNeil, D.J., et al. 1992. Analysis of Streptomyces avermitilis genes required for avermectin biosynthesis utilizing a novel integration vector. *Gene* 111: 61–68.
23. Tong, Y., P. Charusanti, L. Zhang, T. Weber, and S.Y. Lee. 2015. CRISPR-Cas9 based engineering of actinomycetal genomes. *ACS Synthetic Biology* 4: 1020–1029.
24. Tong, Y., et al. 2019. Highly efficient DSB-free base editing for streptomycetes with CRISPR-BEST. *Proceedings of the National academy of Sciences of the United States of America* 116: 20366–20375.
25. Ferrari, S., et al. 2020. Efficient gene editing of human long-term hematopoietic stem cells validated by clonal tracking. *Nature Biotechnology* 38: 1298–1308.